S0-FJZ-459

EDUCATIONAL LEADERSHIP

An Interdisciplinary Perspective

EDUCATIONAL
LEADERSHIP
An Interdisciplinary Perspective

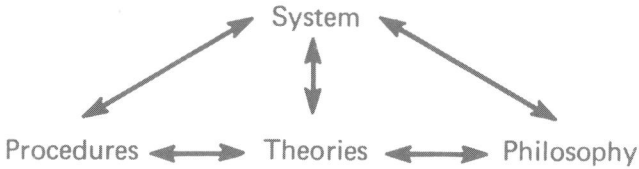

Robert L. Granger
Dean of Education
Newark State College

Intext Educational Publishers
College Division of Intext
Scranton   San Francisco   Toronto   London

ISBN   0-7002-2192-1

Copyright ©, 1971, International Textbook Company

All rights reserved. No part of the material protected by this copyright notice
may be reproduced or utilized in any form or by any means, electronic or
mechanical, including photocopying, recording, or by any informational storage
and retrieval system, without written permission from the copyright owner.
Printed in the United States of America by The Haddon Craftsmen, Inc.,
Scranton, Pennsylvania. Library of Congress Catalog Card Number: 71-160680

LB
2806
.G65

to Virgie Mosby Granger
a lovely wife
and
an invaluable professional associate

199127

Preface

In this era of rapid and complex social change and scientific and technological development, learning experiences for educational administrators must be designed to stimulate an inquiring attitude and to facilitate knowledge and skill capabilities. Specialization in various facets of educational management theory and in instructional and inquiry theory are a part of many effective programs preparing educational leaders of the future. An emphasis on an interdisciplinary and integrative "system" perspective of leadership in educational organizations is also an emerging element in such programs.

It is the purpose of this book to present a relatively simple explanation of system theories and procedures for the intellectually alert educational administrator and the liberally trained educational observer. It will offer a simple introductory translation of the essential components of general and information system theories, methodologies, and technologies, and explain or demonstrate their functional utility and social value for the educational leader.

In order to present the system perspective in its full import, it is necessary to treat the subject of system in a philosophical and theoretical perspective as well as in a practical frame of reference. Any adequate book on systems must discuss the subject in its most universal aspects because it is in the very nature of system ideas and procedures to consider the organismic whole and account for every contingency. A comprehensive system perspective must consider the interrelationships among individual productive events, systems of operations, theories for planning complex operations, and philosophies for determining the values of operations. Therefore this book is intended to be simultaneously theoretical and practical.

The first two parts of this book introduce the reader to the philosophical and theoretical foundations of general system, information, communications, and inquiry concepts and their historical and disciplinary antecedents. It is intended that the reader will perceive from his reading of these chapters that the system perspective is a well-ordered extension of the philosophy and methods of science, including the processes of hypothetical-deductive logical analysis.

Part Three serves primarily a practical or managerial function. A number of standard system techniques applicable to education are identified and described in this section. It attempts to include all system procedures which are relevant to the practice of educational management, particularly those procedures that are currently in use. The reader will find chapters on system analysis, gaming and simulation, scheduling of costly projects, programed budgeting, and operations research.

Parts Four and Five are devoted to the presentation of social system ideas and theories as they apply to the administration of dynamic social organizations and the administration of instruction. In Part Four a chapter on decision-making derived from system theories is one of several chapters developed from interdisciplinary readings. A chapter on instructional program design in Part Five is another practical application of system thinking to educational problems.

Several chapters of the book are particularly relevant to administrators and teachers who are primarily nonmathematical and nonscientific in their social and managerial perspectives, who intuitively reject categorically closed system ideas and applications as useful or desirable instruments for assisting the educational administrator. The chapter on general semantics in Part Two and a chapter entitled *Teaching and Communicating as Art and History* in Part Five will introduce them to the possibility that system theories accommodate open and inductive procedures as well as scientific, systematic, and deterministic processes. There are some notable differences in the assumptions underlying systematic behavior in these diverse classes of systems, however.

In the design and development of the several sections of this book, both a systematic and an open inquiry position are adopted by the author. It is intended that the content development include an assumption of the organismic yet analytic perspective described by Bertrand Russell:

> The painter wants to know what things seem to be, the practical man and philosopher want to know what they are; but the philosopher's wish is stronger than the practical man's and is more troubled by knowledge as to the difficulties of answering the question.
> Russell, *The Problems of Philosophy*, p. 9

Thus it is the desire that this book be both systematic and practical, yet open-ended and transformative. As an explanation of system philosophies, theories, methodologies, and technologies, and their relevance for the educational administrator it is essentially simplified and incomplete. Its value, like that of similar books, lies in its immediate and generative uses.

Robert L. Granger

Union, New Jersey
May, 1971

Contents

I

AN INTERDISCIPLINARY
SYSTEM PERSPECTIVE

Introduction

About two hundred years ago men combined their rather well-developed knowledge of logical analysis and natural science with a method of selective and consecutive guessing and developed a procedure called the experimental or scientific method. Continuous application of this trial and testing method in the solution of human problems has created an industrial-technological revolution. It has also revolutionized mathematics and the physical and biological sciences through the eventual generation of the profound theories and laws of evolution, probability, relativity, and quantum mechanics.

A little more than twenty-five years ago scientists and engineers refined their scientific procedures by combining them efficiently into interdisciplinary theories, in fact, into a metatheory, a well-ordered set of "system" operational rules which is a particularly useful instrument for resolving difficult problems. Among the results of the first system procedures were the rapid and relatively economical development of the technologies of radar, nuclear energy, rocket propulsion, and automation. A common set of theories and methods were employed in all of these complex and expensive technological advances, and technology itself began to make a greater contribution to the advancement of applied science.

Of all of the great human inventions of the last twenty-five years, the system perspective and its methods and techniques, particularly when applied to electronic computation and automation, may have the most far-reaching effect upon mankind. For in systematic automation men have constructed for the first time machines which can regulate other machines or automatically supervise the manipulation of huge amounts of stored symbolic data. Electronic computers as "thinking" machines are very much a reality, and we are now in the third and fourth generations of their evolution and perfection.

At the present time we are beginning to realize the potential and experience the effects of computers and automated technology upon traditional patterns of human production or purpose-oriented activities and thoughts. We believe we are experiencing the dawn of a second technological revolution, a cybercultural revolution, wherein men are making another giant leap forward in the speed and accuracy with which they imagine new goals and apply system analysis methods together with engineering and industrial technology in the process of expediting human progress.

Certainly all educated persons and professional educators, whether scientific or humanitarian in their interest and orientation, are or will be concerned with understanding the fundamental nature of the scientific and system perspectives and processes. They will want to know something of the basic system assumptions and of their structural and functional limits. In order to establish a background for explaining system concepts and operational procedures to the educational leader, this introduction will briefly describe their historical evolution.

The prescientific, historical, or traditional method of human inquiry and purposeful behavior which dominated classical periods of human history is sometimes called the Aristotelian or deductive method. It is a perspective and process in which the new and unexplainable event is related to the natural, the old, the familiar. It was and often still is sensible and economical to begin with the known and systematized in inquiry and relate the mysterious to the intelligible. In substance the Aristotelian, traditional, or historical method of inquiry is a method of deductive analysis based on prior exhaustive observation, enumeration, and classification, and primitive, random or accidental, and inefficient trial and error. It was and is a productive procedure in the simple or primitive stages of human inquiry.

The experimental or imperfect inductive method of human inquiry developed by Francis Bacon, Descartes, Newton, and others deliberately creates and tests hypotheses or probability statements of cause and effect in simulated or staged events in order to expedite or accelerate the more naturalistic, historical, or traditional methods of observation. As B. F. Skinner has said, the experimental method is superior to simple observation and the Aristotelian method because it multiplies accidents, caprice, and curiosity—it accelerates nature in a systematic coverage of the possibilities.

The inductive method of scientific or systematic inquiry, although it advances human knowledge and technology in a step-by-step manner, frequently progresses along certain avenues of investigation with surprising speed, precision, reliability, and economy. System methodology is essentially self-correcting or error-reducing and self-accelerating in any given theoretical or material system. Eventually it makes possible functionally perfect enumeration and operational efficiency, satisfactorily specifying all of the necessary and sufficient properties of elements and operations within extremely complex events and systems. System procedures are particularly effective in solving problems in the "hard" or natural sciences and in material production because, as Norbert Wiener said, natural and physical laws have no deliberate strategy for protecting their secrecy. They are basically rational elements; they are not purposeful, competitive, capricious, or irrational.

Since the inductive method of inquiry relates the deliberately original to the accepted by relating a unique but determinate or reproducible phenomenon to an existing paradigm or theory which is considered tentative rather than absolute, it substantially reverses the deductive method. Inquiry and application now can be more truly experimental rather than predominantly ex post facto.

As new tools adding to the experimental process, system procedures are applicable for effectively and efficiently planning, analyzing, designing, implementing, and evaluating in all academic and practical disciplines. System analysis and redesign operations permit men to accelerate the cycles of (1) creating new images from the systematic review of prior experiences, (2) creating new and more economical material inventions from these synthetic images, and (3) subjecting these above processes to a well-ordered system of interdisciplinary cost-benefit analysis or evaluation.

The importance of man-made machines and instruments should not be discounted in any attempt to understand science and system, because systematic inquiry and system applications constantly extend and revise existing theoretical data and their applications. And, although some prior knowledge is discarded in our fast-moving world, system procedures have a tendency, when not subjected to careful and continuous monitoring, to constantly increase the amount and complexity of knowledge generated. In such a situation computers and data storage systems facilitate both data production and data evaluation just as industrial machines accelerate selective material production. A computer's speed, focal attention, accuracy, and durability of error elimination (they have a prodigious, accessible, and virtually infallible memory) are a tremendous supplement to human mental abilities. It is worthy of note, however, that men must still do the imagining and developing of the symbolic goals, models, and programs which guide their computers, just as the production computers must guide and limit their attached automata.

Through the coupling of the human processes of systematic inquiry, system procedures, and cybernetic technology, we have been able to reduce greatly the lag and cost of transforming human goals into mass production. The lag caused by human perceptual uncertainty and variability is great by comparison. Thus we find that we are living in an era in which men are transforming their environment faster than they can as a class consistently transform many of their human imagery or value systems which monitor their activities. In such a situation human judgment seems to be dependent upon technology rather than vice-versa.

For the persons who wish to accommodate and manage the methods and products of systematic inquiry and technology there are still some traditional postulates from which to draw assurance, however. It remains true that the most acceptable and reliable indicators or predictors of cause-effect relationships permit only retrodictive (past to present) inference, not truly real-time, coincidental, or truly predictive (present to future) inference. Thus systematic inquiry and technology will continue to be subject to or dependent upon human imagination, purposing, continued attention, and a posteriori evaluation and judgment.

In addition to the human control of scientific technology by means of applications of social judgment, there are inherent in human theoretical and systematic processes other built-in controls. Scientists will continue to postulate a relative constancy in the universe of change (Van Dalen and Meyer, pp. 36–37),* wherein experimental change and its standardization in mass production will tend to take place slowly enough for men to draw reliable and valid generalizations about the values involved. Men, in all likelihood, will continue to direct their search primarily toward goals hypothesized to be of some redeeming value. They will continue to work toward the successive elimination of that present in a social or environmental system which is unnecessary or undesirable and toward the discovering of that absent which is necessary and desirable.

In conjunction with postulating a relative constancy and selectivity in the universe of change, scientists and system engineers also advance another related simplifying and controlling postulate. They assert the useful existence of natural or general kinds (ibid., pp. 34–35), wherein men arbitrarily assign to objects and events an identity designation or enumeration, classifying them objectively and judgmentally into units, classes or sets, sets of sets or taxonomies, and functions of sets or operations. In this manner they can construct over a period of time extremely complex taxonomic orders, general theoretical systems, and general systems of values (philosophies), near-perfect universal human induc-

---

* References are to the Bibliography at the end of the text.

tions. Thus all events of significance to men, whether naturally or technologically induced, will continue to be subjected to both a priori and a posteriori intellectual screening and systematic judgment. Within an eclectic system perspective, the traditional, slow historical-philosophical-artistic method of exercising human judgment will interact with and regulate the scientific-experimental-economic method in the evolution of human inquiry, behavior, and wisdom.

Before explaining several of the common system procedures which are attracting considerable interest among educational leaders as well as officials in government and business, it seems useful and desirable to present an overview of the theory of systems. The remainder of this chapter will be directed toward such a purpose. Readers will soon note the organismic and interdisciplinary character of the system frame of reference. It requires such a perspective in order that the metaphor or language comprising it will accommodate the standardized meanings and procedures of interdisciplinary systematic inquiry, various system and information theories, and system technology.

The general theory of systems was first expressed in its eclectic form by Ludwig von Bertalanffy, considered the father of general system theory. As it has been reflected upon, employed and tested as an applied model for constructing human-machine systems, and extended and revised, it has been labeled general system theory, information system theory, cybernetic theory, communications theory, general semantics, and psycholinguistics. With gross generalization it might be said that system and information system theorists and technologists are perhaps more concerned with the orderly description and classification of natural, material, and machine systems, whereas general semanticists, linguists, and inquiry theorists are concerned primarily with the function of languages and codes of sense data or information in systematic human operations.

In reality, however, the heart of all system theories is a theory of information and codes of meaning which will have utility in interpreting human inquiry and direct and indirect applications in man-machine methodologies and technologies.

The remainder of Part One of this book is directed toward the brief presentation of a descriptive general system or information system theory containing the primary system concepts employed extensively in so many applied disciplines today. Part Two is directed toward describing the linguistic factors involved in human inquiry and its systematic applications. If educators wish to fully comprehend the meaning and utility of the system perspective in relation to all problems of educational administration and instruction they will want to read the writings of von

Bertalanffy, Boulding, Wiener, Pask, Ryans and other general and information system theorists and Korzybski, Whorf, Osgood, Miller, Rapoport, Gagné, Chomsky, and other general semanticists and psycholinguists. All of these interdisciplinary systematic philosopher-theorists are attempting to describe and explain human behavior as it interacts with all other natural or man-made systems.

2

General Theories of Systems

A system may be defined as a cohesive collection of items that are dynamically related. The term describes an interrelated network of objects and events or the symbols for such an assembly. Systems are sets of elements or parts which possess some degree of independence or identity but, at the same time, are an integral part of a larger ensemble or whole. In this assembly or network the parts function to produce some process or product which is unique to that particular organismic unit or system.

The idea of hierarchy, of part-whole relationships, is fundamental to the understanding of system theories. Whole systems are composed of parts or subsystems which can be decomposed further into component elements. In the reverse process of assembly or synthesis, subsystems can be combined into systems which ultimately can be enlarged to include the universe. A system perspective is one which perceives an organismic whole and the interrelationships of all subsystem elements. System procedures involve the identification of a primary system or universe, its decomposition, an analysis of its parts, and a reconstruction or redesign of the parts into a new and improved assembly.

Theories of general systems, cybernetics, information systems, and communications provide useful concepts and constructs for the development of an integrated and interdisciplinary system perspective. All of these theories are designed to succinctly but comprehensively present:

The study of networks of interaction
The theories of signals and information
The science of communications and control
Inquiry theory, a theory of human purposive-causative acts and
    actions

9

The multiple goals common to these general theories of systems are to generate:

Theories—these should include universal or cross-disciplinary meanings and definitions.

Applicability—the theories must generate hypotheses that can be tested and used in the real world of men.

The construction of systems—this includes the generation of new and different relationships of elements and meanings.

There are a number of general attributes which all systems have in common. Among these common characteristics are:

1. Order—the natural or typical space-time conditions of the system.
2. Variety—a random or accidental and/or deliberate and purposeful change in internal or external system conditions.
3. Control—the characteristic pattern of order and variation within the system. Affecting this pattern establishes control.

Control is the pattern of connectiveness which maintains the system or transforms it in an orderly or purposive manner. Control methodology is described by specifying the usual and/or an optimum cause and effect pattern or input-output operation within the system. The most reliable or deterministic control method is direct coupling of cause and effect, wherein output is predicted and controlled in a one-to-one relation with input. As input and output connections in systems tend to vary, control is said to be less certain. The system is said to be probabilistic when its patterns of control are uncertain.

Perhaps the simplest and most natural classification typology for distinguishing and categorizing among systems is one in which simple systems are identified as primarily physical, mechanical, living, or symbolic. These more or less natural categories of systems are described as follows:

1. Physical systems—basic or natural systems of inertial energy and matter.
2. Mechanical systems—humanly engineered machines, tools, or technologies.
3. Living systems—living, growing, self-transforming, self-transporting systems.
4. Formal symbolic systems—systems developed by men to representatively reconstruct or preconstruct natural or mechanical systems, living systems, or other symbolic systems. These systems

include the rather precisely coded information systems of verbal communications, logic and mathematics, natural and social laws. They also include certain other formative or primitively informative systems of art, imagination, feeling, and motivation.

Systems have been usefully classified according to their relative complexity. They may be categorized as:

Simple but dynamic
Complex but describable (elaborate but relatively consistent interconnections)
Exceedingly complex (the cause and effect of parts and whole cannot be described precisely or in detail)

It has already been mentioned that systems also can be described by the degree of consistency of interaction between their parts or between one system and another system. Description here can produce at a minimum these categories:

Deterministic systems—parts combine to act in a "perfectly" predictable way
Probabilistic systems—internal and external relationships cannot be predicted with absolute assurance

Theorists have hypothesized that complete knowledge or shared information concerning the universe would void the probabilistic category of systems, but this hypothesis can be regarded as an idealistic and therefore improbable condition rather than a testable condition. It is therefore meaningless in the scientific sense.

Employing the two dimensions of system classification described above it is possible to develop this matrix of system categories (see Gordon Pask in bibliography).

Systems	*Simple*	*Complex*	*Exceedingly Complex*
Deterministic	Hammer	Solar system	None
	House plan	Computers	
		Automation	
Probabilistic	Methods	Profitability	The economy
	engineering	Stimulus-response	Human thought
			A school

It is assumed by system theorists that there are no members of the exceedingly complex class of deterministic systems. All exceedingly complex systems are overwhelmingly probabilistic.

System theorists have created a construct called the Black Box to describe man's ability to observe and control probabilistic elements and systems. They reason that identifiable inputs can be inserted into the system in a controlled manner and apparently related output effects can be observed or measured. But all that transpires in the operating system is not observable, measurable, or immediately understandable. The system can only be perceived as a Black Box. The blackness of the box hides many subtle cause-effect relationships. Only the more obvious input-output relations can be determined with any degree of precision.

In deterministic systems, significant inputs and outputs can be identified rather readily and are therefore subject to relatively exact prediction and deliberate control. In indeterminate or probabilistic systems, the input-output relationship is a homomorphism, a many-to-one transformation incapable of total analysis. Its cause-effect prediction and control is subject to unexpected and uncontrolled variation and the output state of the system may not be what is expected or desired.

Through their efforts to describe, predict, and control the operations of systems, men have discovered that systems can be usefully categorized as *open, closed,* and *isolated. Open* systems are systems which possess networks and network boundaries capable of change or penetration from without or within. They are adaptable and subject to change. They can either receive information or material inputs or initiate information or material output exchanges with other systems. Completely open systems, if such systems could actually exist, would be entirely subject to or dependent upon natural evolution and the nature of events for change, as their self-initiated or regulated behavior would be unceasingly random and incapable of systematic, purposeful, and cumulative modification. They could not stabilize their direction of variation, thereby conserving the good and valuable from the input-output operations and avoiding the recurrence of error (the bad).

Partially open systems have a capacity for cumulative modifiability. They can selectively maintain their new input-output operations and output condition as progressive states of dynamic equilibrium. In such circumstances a relatively constant ratio or balance is maintained among component parts or elements in the system in spite of major change operations. Such systems may be described as possessing the property of self-regulation.

An important condition of partially open or self-regulatory systems is that they display the property of *equifinality*. That is, in such systems identical output conditions appear to be derived from different initial inputs, and identical inputs appear to achieve different results. The logical explanation of this phenomenon is, of course, that the system is extremely complex and all input-output conditions are not accounted for.

The observed input or inputs may be either necessary and/or sufficient in apparently identical systems, but they are not equally or simultaneously so. Obviously there are differing unobservable conditions in the identical systems. It is equally obvious that when equifinal input-output operations take place the order of the several inputs involved is relatively variable or alterable, not linear or sequential. One mediating or dependent causal condition may precede the other or vice versa. This is a common characteristic of extremely complex systems.

Closed systems are naturally or primarily static in their internal relations and in their relations with their external environment. Their boundary structure and general properties are relatively impermeable, irreversible, and unchanging. They lack the capacity for self-initiated behavioral control. Closed systems have a boundary and structural state operationally equivalent to their present functional effectiveness. They are thus substantially determinate and generally irreversible in their direction of change. Machine technology and other algorithmic processes (processes which are determinate and linearly ordered) represent substantially closed systems of control created by men. They always have an extremely limited range of adaptation and order of purpose.

Some physicists specify a special category of closed system, the *isolated* system. In isolated systems there is no transfer of either mass or kinetic energy (inferring no evolution). Isolated systems therefore cannot undergo any definable degree of irreversible or reversible change.

Another important property of systems is the condition of *entropy*. Entropy is a state of a system wherein there is a deterioration in the order and control of that system. It thus involves resisting or unbalancing forces which interfere with the orderly accommodation of new input or external change forces and a subsequent restabilization or balancing of subsystem components as they are assimilated into their new and optimum state of dynamic equilibrium. In communications terminology entropy is referred to as noise, a condition interfering with the accommodation of initial message inputs and their assimilative testing via subsequent feedback transmissions, producing near-perfect fidelity.

Natural or easily identifiable elements of systems are referred to as *objects*. Objects, as component parts of systems, are entities or forces which maintain or conserve their identity and equilibrium during an event. Objects therefore transcend events and systems operations. Objects transformed or transformable during events are called *variables*. Non-transformable objects are called *values,* employing the term value in its material sense.

An *event* is an instance or unit of functional activity or variation in a system, an intersection and transformation of two or more previously independent or semi-independent objects or forces. Events perceived as causative and controllable are called *operations*. It should be noted that events in systems are time, serially, or incrementally ordered in their relationships. A single event represents simultaneity of space-time-function. One event in a system can be related to another only by comparing their incremental dimensions. Objects and events are usefully perceived as simultaneously spatially ordered as well as time-ordered particulars or subsystems within a larger universe. This hierarchically ordered spatial relationship, the part-whole relationship, also is essential for comprehending the functions of possibility, probability, and economy in the control of systematic operations.

During an event in a system one or more objects or forces are accidentally or purposefully brought together as inputs and are in some unknown and unobservable manner transformed or affected by this intersection. In purposeful human behavior, men attempt to create desired output states in objects or systems by regulating the input resources and conditions during events. They control, measure and observe their quality, number, and space-time order. Later, at some arbitrarily or naturally designated moment, the systematic inquirer or operator again carefully regulates, observes, and measures the output state of all of the original objects and forces involved. This is the method of experimental science and research. After a series of such trials, the evolving method of science usually confirms the possibility and probability of causative control.

CONTROLLED EVENTS AND OPERATIONS IN SYSTEMS

The structure of an event in a system can be explained more readily if it is pictured symbolically. Standard system symbols and set relation symbols are employed in Fig. 1 below to depict an event, a single episode in the structure and operation of a system.

There must exist prior to an event at least two independent objects or forces (subsystems) and the space-time possibility of their functional coincidence and transformation. Next must follow an instance of interaction. This is usually followed by a subsequent partial separation. Under controlled conditions the particular instance, order, and degree of the inputs will be deliberately and carefully regulated. Results of event transformations or operations are predetermined to a great extent by the natural boundaries and properties of the objects and forces involved, but precise human regulation of all input conditions can accomplish wondrous things.

A transformation (step 3) occurring during an event may resolve or translate itself in a number of ways (6, 7, 8). It will display during the event

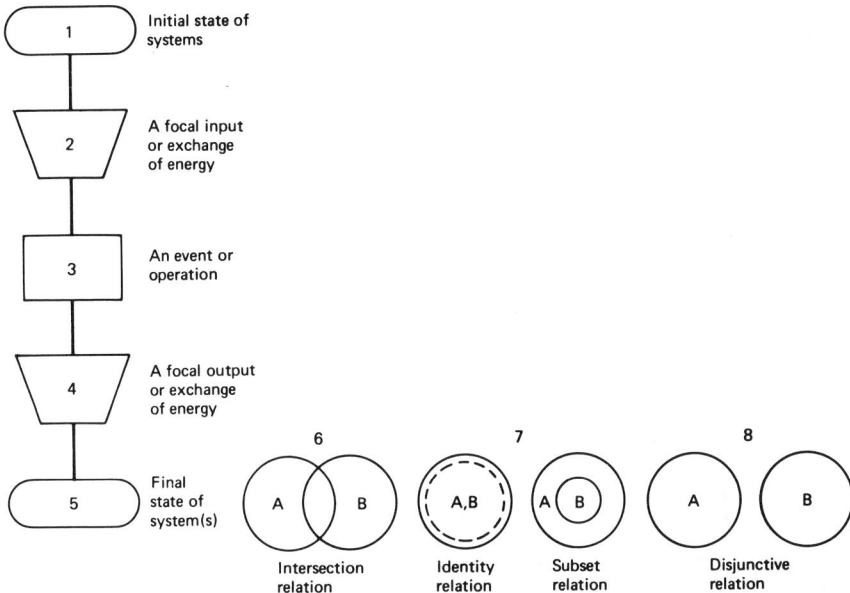

Fig. 1. An event in a system.

a partially unobservable, unexplainable "Black-Box" behavior in the fusion of primary and secondary input systems now said to be in conjunction (6). At the termination of an event, determined by the arbitrary or natural establishment of a final checkpoint or controlled reseparation of sub-elements, the subsystems involved may assume four possible relation-ships. Some unknown or "Black-Box" effect of the conjunction will con-tinue to exist although the degree of its importance may be negligible (6). In addition, the result of the transformation may produce a single new and unique form (7), either adopting the image of the primary sys-tem (the identity relation) or assuming an additive or cumulative relation (a new identity or subset relation). Another possibility is that both sys-tems will become completely independent once again, exhibiting an immediate or long-range effect which can range from negligible to great (8), a disjunctive relation. It is important to note that both the primary and secondary input systems experiencing an event, transformation, or operation are always affected irreversibly or incrementally as well as reversibly, although the degree of these effects may be neither equal nor cumulatively important.

Careful and measured comparisons of input and output states of all system components before and after an event will reveal the immediate significance or act meaning of the event. Subsequent evaluations are necessary to measure the systematic or persisting effects, the action meaning of events upon the systems involved. Effects of some systematic operations may not be immediately detectable.

It is much easier for men to observe, describe, and predict behavior in extremely complex systems than it is to control that behavior. In all systems it is much easier for men to allow a system to continue its present or natural state rather than change it. This is the path of least resistance, but it may not be the optimum path from the human point of view. However, forecasting is not the major value of scientific or system procedures. The prime goal is to effect one or more specified, limited, causative transformations or controlled operations in a system.

Incidentally, it is much easier to produce material products and factual information in systems than it is to assess their ultimate or even their relative value. It is highly significant that scientific technology per se can be exploited by the ignorant (those not possessing or understanding the long-range or ultimate significance or implication of causal events) as well as by the wise (those who subject events to the evolving and dynamic value systems of recorded human affairs). Science as technology is event-based or causal-analytic. Science as systematic human philosophy and history must supplement and complement experimentation and technology if it is to be system-based or causal-syncretic.

A careful sequencing of all experimental trials and subsequent tests of operational technologies for the purpose of selecting an optimum standardized production technology is the system process. Sequential experimentation systematically regulates results in a heuristic iteration, a search or discovery system, that eventually almost guarantees the generation of new knowledge and efficient and economical technological control in very complex determinate systems.

When system procedures are eventually mechanized and mathematically standardized so that all operations are prescribed (closed or accounted for in advance), the resulting plan or program can be considered an algorithm; its ordered inputs and operations automatically assure the desired output or result. Technology is the material embodiment of humanly designed algorithms subjected to additional economic regulation. Repetitive or systematic mass production of standardized unit outputs is the typical goal or product of a closed system of technological operations directed by completely prescribed programs.

Continuous non-standardized self-correcting human heuristic capabilities are a necessary supplement of technology, however. In understanding and controlling a new or complex environment, heuristic or search-oriented methods are always needed. In such instances the systems involved are said to be complex and ill-defined (1) because desired input and output states are uncertain or variable, and (2) because alternative methods for achieving these states exist and their comparative value or efficiency is as yet unknown. In a very real and practical sense, technology and operational efficiency are possible only in systems where men have considerable prior knowledge, experience, and commitment of interest and resources.

Although system procedures have proved to be most applicable for advancing the natural sciences and for optimizing mass material production and distribution, they are applicable in some degree to the solution of any unique or persisting social problem, including education. Unfortunately the use of system methodology and technology for predicting and manipulating human events is often dysfunctional, premature, and inappropriate, primarily because of the microscopic perspective employed to guide the analysis and redesign of the system under consideration.

CYBERNETIC SYSTEMS—SYSTEMS WITH DELIBERATE OR ALTERABLE SELF-CONTROL

The capacity for control is impossible and nonexistent in any system that is absolutely invariant or closed or absolutely or randomly varying or open. But all identifiable systems which can be said to experience systematic operations are by definition neither completely closed or open but only alternately and relatively so.

That system which is of particular interest to educators, behavioral scientists, and system analysts is the self-regulating, "thinking" or intelligent system. Such a system is called cybernetic or self-governing. A cybernetic system is a system which possesses a special control or information network which sends feedback data regarding output conditions of systems to a control center which compares it with programed performance expectations. Depending on the results of the comparison, the control center then sends out data which determines the further operation of the system.

Extremely complex cybernetic systems are capable of arbitrarily or unnaturally altering the nature, direction, or degree of their adaptive (reactive) or self-initiating (proactive) behavior; by means of the feedback network which monitors observed or measured effects they can establish an incremental or progressive pattern of trials which give sequence and direction to subsequent behavioral modifications. Such sophisticated cybernetic systems alternate between patterns of forward motion, search, or induction and patterns of conservation, recycling, deduction, and integration. They vary between states which are spontaneous, autocatalytic, independent, event-determining, and open and states which are reactive, dependent, other-controlled, adaptive, system-determined, and closed.

In automation men invented a system of feedback controls for machines. These controls measure and regulate the form and order of mechanical operations, monitoring all of the significant system processes as they interrupt and transform the input or natural state of some element, moving it along in a linear manner through series of appropriately ordered operations, each of which transforms it progressively and incrementally into a state nearer its predetermined output or product objective. As the production progresses, the feedback network inspects all objects following each operation, determining whether conditions are favorable for the next event to occur. Each event is therefore dependent upon prior events and conditions. The feedback control system, frequently merely a thermostat or switch, is the brain of the "thinking" machine. The first such machines were deterministic idiots, to say the least.

The capacity of mechanical brains or programed feedback control systems was greatly enriched by the development of the electronic computer which can follow slavishly quite complicated prescribed programs or directional codes, making all necessary decisions or measuring calculations required in the monitoring process. Computers are the ideal "cheerful robot" for supervising or controlling technological operations where the ordered sequence of input and output states of the system can be completely predefined and controlled. Automated electronic computer control systems can follow a prescribed code or program precisely and accurately with almost unlimited speed. All restrictions on speed (rate of "thinking") in computer systems engaging in simulation (the processing of formal symbolic data only) are determined primarily by the mechanical limits of the input and output devices. In automated manufacture the rate of production is determined by physical-mechanical operations also, except when the computer is deliberately used to regulate the rate of progress.

As the speed and the program and data storage capacity of computers have increased, they have gained the ability to make more complex decisions, searching or branching and comparing among a limited number of predetermined alternatives in order to improve the quality of each successive decision. By means of heuristic programing and iteration they can even search in some consecutive or random patterns among alternatives not specifically identified by the programer. Yet computer processes are fundamentally algorithmic, reconstructed, standardized, and closed operations. The original data input, the directional order and objectives, the limits of search, and the nature or state of all input and output data must be entirely predetermined. Computers cannot simultaneously or alternately modify, significantly reorder, or reverse their output goals and operations, their message events and predetermined coming systems. They lack the capacity arbitrarily to introduce new criteria for wise judgment. Such closed cybernetic systems depend upon men for the original order and definition of the problems or values they consider, for the focus of their rather limited attention, for the reprograming of operations when new inputs are introduced, and for both the starting and stopping of their operations.

Computers and electronic data processing machines greatly expedite the use of formal symbolic systems in prestaging, planning, programing, and simulating complex series of operations, however. With their extensive and infallible memory capabilities, their speed of data access and manipulation, and their precision and accuracy of operation, cybernetically guided technological machines are markedly accelerating human capabilities in planning and in production. The human inquiry and application process is sharply advanced.

Understandably, as men have progressively refined their planning and programing methodologies and their automated technologies, they have employed the theoretical models in these processes to describe the thought processes of human beings and other living creatures. Thus the brain of animals is regarded as a cybernetic control center, and human inquiry and behavior is explained in the system metaphor.

Of all known systems, man is perceived to be that cybernetic system capable of regulating himself and all other known systems (physical, mechanical, living, and formal symbolic systems) most effectively. He has a special range of dynamics, special capacities for cumulative modifiability, for adaptability, flexibility or reversibility, for caprice, for imagination, growth, and change, for decay, regeneration and repair, for speed and extent of response. Man is the most complex, capable, and sensitive of systems. He can therefore exhibit the greatest departure of output from prior explicitly programed norms intended to evoke a specific adjustive output or response. In other words, man is the most creative, mentally proactive, and arbitrary of systems and therefore the most difficult to control.

Of particular importance in understanding humans as a particular class of cybernetic systems are the complex and flexible properties of human imagination, adaptive memory, and motivation and their use of formal but flexible systems of communications. The special capacities are all essential in the creation of present human technologies and in their further development. A more lengthy description of human imagination and applied intelligence will follow in another chapter. Formal symbolic systems employed in systematic communications will be discussed elsewhere also. The fundamental or basic properties of cybernetic systems and information systems are the focus of this chapter.

The problem of prediction and, to a lesser extent, control in natural systems, machine systems, and mechanical cybernetic systems or robots is essentially deterministic, therefore relatively easy and accurate. Here the scientist is concerned with causation that operates genetically (in a generally linear, incremental, and well-ordered sequence). Thus, a careful predetermination of input conditions and a predetermination of specified feasible output conditions pretty well closes or determines the system.

Operations researchers and applied scientists, engineers, and technologists are quite successful in describing, controlling, and reconstructing simple and complex deterministic systems. Using applied system techniques in deterministic systems, men can plan, analyze, design, operationalize, and evaluate normative or standardized production systems or machines, improving the possibility, probability, and economy of their goal attainment. System theories and technologies are the most immediately successful when they are applied to such environmental or systematic conditions.

Yet the advanced system theorist or analyst is concerned with improving his knowledge and control of exceedingly complex and ill-defined probabilistic systems, systems significantly affected by human behavior. He has this focal interest in company with knowledge theorists, behavioral scientists, and educators.

In self-controlling, self-transforming human systems, methodological iteration is intuitive, heuristic, irregular, and reversible. Self-inductions and therefore systematic operations are extremely complex and uncertain. Imaginative and behavioral inductions may be pluralistic; they may leap gaps; they may be equifinal, substantially reversing prior input-output relationships or experiences. Humans, responding to idealizations and imagination, truly operate from the future to the present rather than vice-versa. Thus, intelligent, informative, retrodictive behavior based on past events and experiences, may be substantially reversed, ignored, or discarded (forgotten). Caprice may result. Relatively great freedom of intuition, will, and choice may be exercised in an irrational confrontation of great odds. Nature may be significantly transformed.

External regulation of healthy, active, curious human and social systems by other humans is thus very difficult and never complete or failsafe. Regulation of cheerful mechanical robots is one thing; regulation of creative, intelligent, educated, and wise human beings is another. Relatively constant behavior in creative and intelligent human societies can be accomplished only by keeping the system simultaneously and/or alternately open and integrated, free and interdependent.

It is just this balance between stability and change, conservativism and liberalism, order and variety, that is the problem of the social leader or theorist. Educators, behavioral scientists, system analysts, and systematic philosophers of all kinds are focally concerned with these extremely dynamic yet delicately balanced systems. Their system of rules or science must be carefully open, yet intelligently closed, just as is the case with the natural scientist. Yet they must accommodate and assimilate extremely complex patterns of operations and events into a complex theoretical and value system in a progressive and optimizing manner.

An awareness of the constant growth and change in human societies as well as in nature is a prerequisite for understanding individual human beings and social systems. Social theorists have generally concluded that a sense of this general condition and an opportunity for participating in the creative decision-making, constructive action, and normative evaluation, of interdependent social and social-environmental systems are necessary components of healthy and continuously productive systems under human control. This sense of awareness and opportunity must exist for all who possess the desire to influence the system if they are to be expected to cooperate voluntarily and continually. In such circumstances efficient communications and reasonably equitable distribution of the system's input resources (ideal as well as material), its effective methodologies of inquiry and technology, and its output products or optimizing conditions must occur. Otherwise social frustration, unproductive competition, and conflict will prevail over satisfaction and cooperation. In addition, humans are becoming acutely aware of the necessity for achieving and maintaining an integrated and balanced relationship with their physical environment or they eventually will be negatively affected or possibly destroyed by this lack of concern.

Although system theories and metaphors attempt to account for and explain the special human capacities of creativity, intellect, memory, and symbolic communication, they also explain the nature and existence of ever-present environmental restrictions and limits which order and simplify human affairs. These material and intellectual realities are accounted for in the postulates of limited induction, relative stability and change, natural kinds, and in other laws of nature and of science. At any one instant, in any single operation or system, human capabilities are incomplete and therefore restricted. And they are always dependent to some extent upon the prior nature and disposition of all the natural elements comprising the environment. There is always a lack of information, an unmanageable complexity of data, and a lack of immediate access to the relevant data. Human prediction and its application in control operations is always substantially retrodictive or incomplete. In any event one might readily predict what will happen under certain conditions, but one cannot simultaneously or instantaneously predict and/or establish the existence of these conditions and the effect of the action. Thus we come to a brief discussion of the limits of system control based upon retrodictive methods and probability theories.

Probability is defined as the relative frequency with which a particular event will occur in a class (an abstract, standardized, normalized system or class) of events or conditions. The probability of a unique event is meaningless; unique events are only possible, absolute, or categorical. No event is probable intrinsically but only in terms of its membership in certain classes, sets, or series of events or operations. Probability is the relation or ratio of favorable or specified events to possible events. It is an incremental or optimizing class or set of events, deduced or defined by prior retrodictive experience and evidence, which has greater credibility than other categories in the range of possibilities but less credibility than the sum of all independent probabilities (the domain of possibilities). Probabilistic events and systems therefore exist only in the abstract or non-particular, in circumstances where they are alterable and where their alteration may effect changes in subsequent conditions or states. Possibility, probability, and perfectibility are possible only in semi-open systems.

When a unique event or operation occurs in any system the final output state always represents to some degree a new object or material equilibrium. It therefore is generative, capable of further modifying the domain of possibilities and range of probabilities of the system. Possibility and probability are therefore both relative in the evolving metaphysics or philosophy of systems and operations, in the theories and strategies of interdisciplinary science. Possibility is absolute only in the present or concrete; probability is absolute only in the past and future abstract.

In the evolving theory of systems with known and unknown patterns of irreversible and reversible operation, there are no total absolutes, but there are systems of practical possibilities and abstract probabilities which simplify or usefully order and control human and natural affairs. The appropriate and alternate use of these practical systems of concrete and abstract induction and deduction is the system approach in living. As Norbert Wiener said, "The task of living is to swim upstream against the current of entropy."

In addition to Wiener, John Dewey, Bertrand Russell, and Albert Einstein were among the first great intellectuals and systematic theorists to recognize the dynamic yet incomplete nature of systematic human inquiry and experimentation processes. They recognized that no two events are absolutely identical, that total reconstruction or prediction of events in extremely complex systems is impossible. Yet these men sensed the potential impact, near-universal applicability, and probable acceleration of human control over nature brought about by the man-made invention and extension of inquiry and technological systems.

For, in spite of the fact that systematic methods and models are most effective and precise when they are applied to determinate systems, all behavioral scientists, including educators, can and must employ and

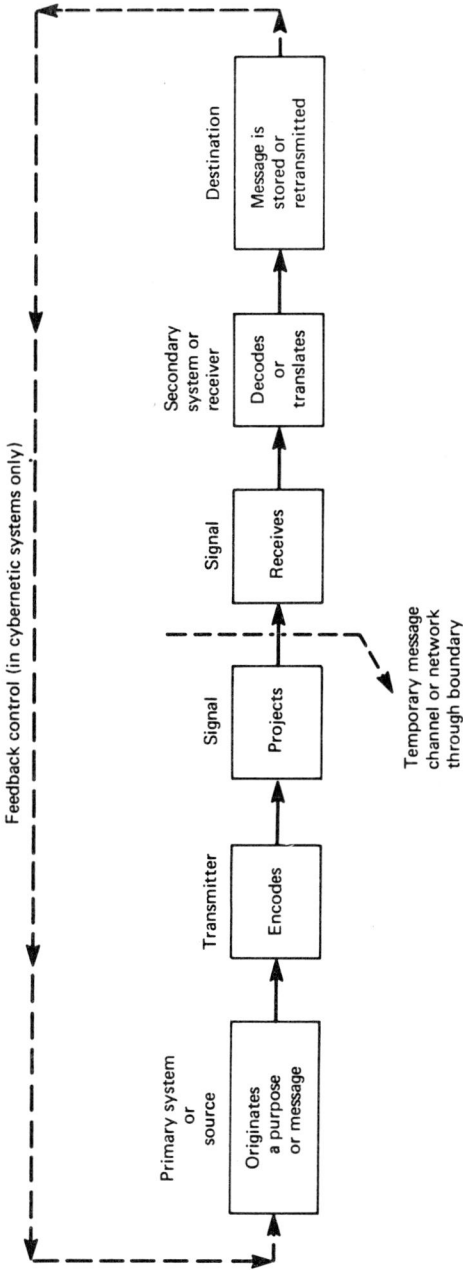

FIG. 2. The communicative process.

adapt them as they develop human methodologies and explain and control systems involving human beings. Only in this manner can they advance their understanding and control of extremely complex systems.

Although progress will be slow and efforts will always be less precise in systems involving humans, the step-by-step methods of science and system are practically assured of advancing and expediting the art and science of education. This is highly probable if not almost inevitable. There is no other strategy available to man which offers anywhere near this degree of possibility, probability, and economy.

A COMMUNICATIONS MODEL

A meaningful perspective of the operation of a self-regulating or cybernetic control system can be readily developed through the use of the information or communications model. This linear flow diagram of a communications or mutual interaction process showing the feedback iteration or cycle is a simple metaphor, model, or analogue of any self-regulating or intelligent system operation.

In the communications network there must exist prior to a communications operation or event:

1. Two semi-independent elements, a primary system or source and a secondary system or destination, existing in a state or channel (possessing simultaneity in space-time) permitting their intersection and interaction. An interaction must occur.
2. A previously distributed code or set of common conditions (shared natural or formal components) which will facilitate the sending and capturing (exchange) of whatever message is communicated. A message may be primarily a physical act or operation transmitting energy which acts on mass or mass itself, or it may be in a symbolic form, transmitting energy as information or coded sense, sound, sight, formal symbols, etc., acting through and upon cognitive or cybernetic control systems.
3. An energy or information imbalance or difference between source and destination. This may be in the form of a need or purpose (explicit or implicit), precipitating action in the self-initiating or primary source element. This imbalance determines the subject-object relationship, the direction, intent, and extent of the energy flow, the meaning of the message.

The unique character of information input content contained in any single message or operation is that previously undistributed element(s) prior to the event which is distributed (transferred, changed, redistributed) following the event. The remaining content in the message is the previously distributed translation code which remains systematic, redundant, stable, and entropic. Expressed another way, the content of a message reflects determinants, which "load" the operational probabilities of an interacting system during an event. They generate a significant point of action or decision, simultaneously creating "rules of interaction" and their effect.

Conditions existing prior to the establishment of communications or interaction between a source and a destination include the previously distributed or transmitted common elements or code and at least one uncommon or formative element which is to be initially induced, transforming the receiving or destination system (and, incidentally, also the primary source). There is a definite direction and momentum of flow for each explicit message or event in a communications network. Any system therefore has these sets of elements—distributed elements, undistributed elements, and a functional and/or functioning transportation network. It is important to note that simultaneity or coincidence in space-time-function is a requisite for any communication or operation, determining both its possibility and its retrodictive probability.

Characteristic of all communications processes is the presence of noise or entropy in the system. Entropy is that proportion of a message which inhibits its translation (initial transformation and feedback transmission) with perfect fidelity. It is comprised of all that is interfering or resistant in the network to the particular communications act and its persisting effect or action. A primitive or simple communications operation is one in which any system sends a message transforming the form and structure of the receiving system. A meaningful or intelligent communications event requires, in addition, a previously distributed code and one or more feedback circuits which will carry a return message containing information which will update the code or memory of the sending system. Shared communication between two intelligent systems involves a high fidelity translation wherein the codes of the two systems are updated in a similar manner. The linguistic purist should note, however, that the primary question of meaningful communication or action is not which code should be used but which code is actually employed.

A considerable degree of redundancy is obviously necessary in all meaningful or controlled communications events, especially if they are extremely complex or involve two or more dynamic self-regulating systems. That proportion of a message which is distributed prior to an event (the code) is, for a particular communication or message, redundant. Redundancy can be defined as that proportion of a message, which, if lost, would leave the message essentially complete. Redundancy is said to be negatively related to operational efficiency; relative efficiency or functional control of a simple communications event may be described as the degree to which it approaches one minus redundancy (1 − redundancy), wherein the intent of the source transforms veridically (immediately, precisely, and fully) the receiving system. A perfect communications operation would be categorically certain or determinate and the source desire and receiver accomplishment would be identical and simultaneous. This is about how a computer sends its very focal but highly restricted type of messages.

In communications operations between two intelligent cybernetic systems, each message may be considered as an empty or incomplete form, to which the transmitter and the recipient may attribute the same and/or a different meaning depending upon the code that is applied. In order for such a message to be mutually meaningful it must be transmitted in a context which is highly redundant or consistent with previously distributed codes, cognitive maps, and value system boundaries in both the sender and receiver systems.

In all highly rational communications events and systematic codes under deliberate human control, there is an incremental and hierarchical value order among interacting input-output messages or forces. The exact order of these values or determining forces is dependent upon both natural conditions and human values guiding nature. In a rational communications operation the subject or primary system initiates an operation or communication intended to affect the object or receiver system. In this interaction, the irreversible characteristics of the natural forces involved plus the irreversible characteristics of predetermined human purposing constrain the interaction, shaping, binding, directing, structuring, and limiting the communications event. In any rational or carefully controlled or precoded event, low order input conditions or values are deliberately consumed or expended in the activity of disordering, remapping, and reordering the initial state of the primary or secondary subsystems at the time or point of input so that higher order precious values or output objectives are achieved.

Implicit in most human communications events or operations is the subjective or precious value of the human elements involved. Yet when men set in action systematic communications operations they specifically establish new or additional ordered relationships among all elements in the system.

Both explicit human objectives (messages) and implicit human goals (codes) serve to shape the behavior of human communications operations. Ordered values in human messages and codes serve to constrain, obligate, cohere, cement, engage, compel, and commit elements in any planned event or sequence of events. Negative elements or values primarily serve to disengage or repel. Positive values represent idealized output states, focal, intense, and protected factors which serve as positive determiners in guiding the functioning of intelligent or rational communication behavior.

As men become more skillful in developing sophisticated and elaborate communication and meaning codes, they are effectively learning to identify long-range or complex objectives and plan the series of operations necessary to systematically achieve these objectives. A well-ordered system or code of personal and social values is necessary to facilitate the orderly and consistent regulation of human communications and inquiry, and their applications in operational methodology and technology. A functionally simple yet sufficiently eclectic and flexible social value system or code will become increasingly necessary as men progressively accelerate their control over natural forces. Thus men must also accelerate their ability to plan more satisfactorily, communicate more effectively and efficiently, and evaluate more carefully.

During any humanly controlled communications event there is always functioning a subjective or implicit value order or code and an objective or explicit statement of input-output relationships (a message) in both the sending and receiving systems. The general construct *what* is usually employed to specify in a neutral, objective, or factual manner all value objects or properties contained in the communications operation or event. Among the universe set of *what* conditions or variables frequently transmitted in a communications event are the following:

> *What* refers to any and all systems elements or processes whose value determiners are unordered or disordered.
>
> *Who* and *whom* are code determiners which indicate a humanitarian subject-object relationship in the communications system.
>
> *When* is a code which establishes a time-space order or boundary.
>
> *Where* is a code which establishes a space-time order or boundary.
>
> *Why* indicates message code elements which identify subjective and objective purposes or expected output conditions.
>
> *How* relates to coded information specifying input-output or cause-effect operations.
>
> *How much* is a code which numerically indexes or scales all objects and forces involved in the communications event.

In human affairs or humanly controlled communications and operations the *who, why,* and *whom* variables are the subjective or idealized human determiners or values frequently held in highest priority in thought and action. Where nature and technology prevail or where men act with pure rationality all of the *whats* are considered variable and natural forces and/or representative laws determine the *how, when, where,* and *how much* conditions. Thus there is a fundamental difference in communications systems under purposeful or intelligent human control and those under natural control only. The former communications operations and their component systems are infinitely more complex and indeterminate.

Human communication is facilitated by the use of precisely predefined or predistributed mathematical and verbal codes. This is particularly true when implicit human value codes are also distributed or shared in advance of a communications act. Under such conditions messages can be sent with the expectation that they will receive high-fidelity decoding.

Feedback is necessary in communications if a sender wishes to fully monitor and carefully control the effect of a communications event in a network or system, however. Continuous feedback will permit the sending or receiving of additional cues or context to correct the previously distributed code and improve the subsequent decoding of a significant message. Either positive or negative feedback is more useful than no feedback in communications control. Negative feedback changes the behavior of the transmitter. It counteracts continuation of the present output effort. Positive feedback functions to habituate or stabilize communications behavior. Positive feedback plus incremental iteration may produce deviation amplification or cumulative modifiability in a transmitter or primary system.

In communications between two intelligent and dynamic systems, meaningful communications and subsequent operations are monitored or controlled through dual feedback systems in the sending and receiving systems. If the receiving system is transformed in a way approximating the intent of the sender, the communication is said to be successful from the transmitter's point of view. The optimum state of affairs in human communications is achieved when the receiver via his own feedback system reaches a similar concluding judgmental state. If the feedback received by two communicating individuals is mutual or similar it is reinforcing or complementary; the past action is mutually sustained or resisted. If the feedback is negative in one system and positive in the other, it is interfering and entropic and will result in subsequent counteracting and competing actions.

Social communications may have as a specific purpose immediate retransmission (an instantaneous change in the receiver's behavior). Often the intent of a message is entirely event-based, directed toward changing the behavior of the receiver immediately and temporarily (reversibly). At other times the sender may intend to effect a systematic, irreversible, or long-term transformation in the receiving system. He may wish to change both immediately and permanently the behavior and cognitive map or code of the receiver. Upon occasion a message is sent with the sole intent of immediate storage; no immediate response is desired, but the message is intended to effect a change in the receiver's code only and be accessible for later recall.

In purposeful communication, including education, direct verbal exhortation, exhortative action, and sometimes indirect vicarious esthetic experiences are designed to have immediate effect upon students. Direct experiences usually tend to have both immediate and long-range impact upon a receiver's actions and memory code, as do powerful emotional experiences. Formal explanatory and designative communications are intensionally a very indirect and ordered means of transmitting information. They are usually employed when a complex message is transmitted so as to have a minimum of immediate impact upon the receiver's behavior. Their function is to establish in the stored symbolic code an orderly or linear program for directing subsequent series of actions.

Ordinary verbal discourse often has intensional or artistic connotation as well as objective designation built into its code. It is not designed for precise decoding alone, because it is designed to serve in both objective and subjective communication. As it is highly flexible and adaptable, it can be employed alternately for transmitting formative and informative content, intensional and extensional evidence, artistic and scientific meaning. Each individual occasion of use of verbal communication requires considerable adaptation of the code in order to facilitate the proper or intended interpretation.

Designative definitions of words, logical and mathematical symbols, tend to formalize the structuring of precise human verbal communications because they firmly fix the code used in interpreting such messages. Subjective or artistic messages, full of implicit meanings are purposefully or necessarily ambiguous. An artistic or esthetic message is primitive, subjective, open, and slow; its coding is ambiguous and undistributed. It is often deliberately designated to present content which encourages or permits many different decodings, but its subjectivity may be merely a reflection of the fact that it is original and unique (not fully meaningful) even in the mind of the sender.

SUMMARY

Perhaps the most effective way to conclude this introduction to basic concepts and theories of general and information systems is to provide a list of important terms and an acceptable definition reflecting their meaning. A list of basic system concepts follows:

Algorithms—procedures or operations so prescribed that they automatically guarantee the solution to problems through their repetitive operation.

Closed system—a system characterized by external and internal stability; its input-output relationships remain balanced and stereotyped or standardized.

Computer—a machine or system of machines capable of calculating or performing a sequence of arithmetical and/or logical operations in accordance with completely preprogramed instructions.

Control—the characteristic pattern, connectiveness, or meaning of input-output relationships in a system; control is exercised when inputs into a system are regulated, thereby affecting outputs.

Cue—partial information assisting in the determination of the meaning of a message code.

Cybernetic system—systems with an information or signal network which regulates its basic physical-mechanical operations.

Decision—the reduction of an information input to one of a set of two or more identities and/or differences; selection of one from among two or more choices or options.

Entropy—energy or information input-output behavior that is non-utilitarian and/or negatively interfering.

Equifinality—a property of a system which permits different results from similar inputs and similar results from alternate inputs.

Event—an instance or unit of functional or operational activity, variation, or transformation in a system.

Feedback—a return communication or reaction to information processing behavior; a control process transmitting a portion of the output behavior of a producing system back to the control or decision center or property in the same system.

Heuristics—techniques or procedures for varying search, choice, or discovery operations to permit the progressive resolution of difficult problems; if the heuristics are applicable they will provide a short-cut to a goal; heuristic methods cannot guarantee solution to complex problems.

Human and humanoid systems—self-regulating, self-transporting systems which simultaneously or alternately regulate or modify their output behavior parameters including a predefined program and their input selection processes; these systems are flexible, adaptive, and modifiable by both self-induced and environmental inputs.

Information—energy or sense data from one system which is communicated as random or formal code to another system, affecting its subsequent action.

Information flow—the direction of movement of information or energy in a system.

Information processing—selection and treatment of available information inputs in order to ascertain meaning, establish order, or regulate the attainment of a predetermined operational objective in a system.

Information reception—recovery of signals or information out of an external system or world of noise; this recovery is expedited by a code which essentially requires an agreed upon or stipulated meaning; or it requires a process of searching for such meaning.

Input—selection or acceptance by one system of an energy or material output of another system.

Intelligent systems—systems which employ predetermined and specified operational criteria or codes to function as output parameter useful in regulating present conditions and on-going operations.

Message—a succession of signals or signs having a code or alphabet which can be classified according to similarities and differences, thus possessing potential meaning.

Model—a facsimile capturing the essence of one or more characteristics of an original; models may be physical or iconic, conceptual, and mathematical.

Negative feedback—corrective or error-reducing information return, terminating or redirecting present behavior output.

Object—an element, component, subsystem, entity, or force which maintains or conserves its identity or dynamic equilibrium during an event or operation of a system.

Open system—a system whose boundary networks are readily affected by inputs from other systems, consequently changing internal subsystem conditions; open systems are characterized by incremental change states.

Operand—a secondary element or subsystem which is the focus or target of a deliberate transforming operation in a system.

Operation—a deliberately induced input-output transformation or event in a system.

Operator—the primary initiator or regulator of input in a system; the element or subsystem in control.

Order—the natural or typical state of a system.

Output—a unique property, state, or condition produced by an operation in a system; the product of a process, effect of a cause.

Positive feedback—information return reinforcing or sustaining the system's positive and progressive operation, often resulting in deviation amplification, habituation, and standardization.

Program—a set of precise and ordered instructions (a code) so arranged that it serves as a detailed and objective plan for guiding and regulating a series of operations.

Simulation—model codes or facsimiles representative of the essential elements of the systems being simulated; simulation may involve structure, function, or both structure and function.

Steady state—the characteristic, ordered, or habituated state or dynamic equilibrium of a system.

System—a cohesive collection of elements that are dynamically related; a network of object and event states or the symbols for such an assembly.

Variety—accidental or random change or a deliberate and purposeful change in the state of a system.

II

SYMBOLS FOR THE EFFECTIVE AND EFFICIENT REPRESENTATION OF SYSTEMS

The content of Part Two is included in this book to introduce, explain briefly, and demonstrate the function of linguistic symbols in general communications and in systematic inquiry and planning. The degree of adjustment of our most valuable inquiry tool, ordinary language, that is necessary to permit its effective and efficient use in precise or highly rational thought and communication is described. Such an adjustment demonstrates the order of adaptation employed in converting human perceptual and motor skills into completely rational thought and reconstructive behavior—behavior similar to that represented in the cheerful robot or automated machine.

There are several bases upon which this section is justified as a logical component of a book on the system perspective and its applications to education. Perhaps the primary reason is that it is included to reinforce the reader's understanding that various system and information theories and techniques all have their foundation in general inquiry theory or epistemology and in the philosophy and rules of procedure called the scientific method. General system theory and information system theory introduced in Part One are in a sense special metaphors of inquiry theory. General semantics and psycholinguistics are another set of related metaphors which describe the same organismic human-environmental system in a relatively similar yet slightly different way. In full perspective they all attempt to describe a general inquiry method, one involving human imagination and systematic methods and technologies of inquiry.

If the system perspective is perceived as an interdisciplinary application of the philosophy and methods of science, perhaps it will be more readily understood and accepted. And perhaps the students of system theories will not detour into the construction of systems of neologisms or specially developed word constructs that substantially duplicate acceptable terms with identical meanings that are already in standard usage in inquiry literature.

The inclusion of Part Two also is justified via the belief that perhaps nowhere else other than in the study of linguistics is the difference between general intuitive-evaluative human behavior and systematic or rational "machine" behavior demonstrated so clearly. The difference between quasi-rational humans and omni-rational computers is reflected in the content and organization of the languages used in these divergent systems.

Finally, this section is included for yet another reason. If professors of educational administration want to update their professional preparation programs effectively, they will make them more truly comprehensive and interdisciplinary. Professional leadership programs need to be based in a systematic perspective which is fully eclectic and integrated and which includes and reflects experiences which are significantly relevant to our modern social and technological conditions. An understanding of the operational logic employed in system analysis and systematic decision-making and an understanding of psycholinguistics and the function of rhetoric and persuasive communication in political-social leadership situations and in instruction seem to be two primary components of an adequate administrative training program in today's world. The system perspective clearly points to these disciplines as two extremely important but perhaps not fully appreciated facilitators of effective educational leadership and communication.

3

General Semantics and Psycholinguistics

Symbols represent something else. When organized into customary and/or designative codes they serve the purpose of classifying all known elements of existence, the relations among these elements, and the operations men employ to affect or control these elements and understand them. Thus they become a complete yet relatively simple code for explaining reality, simulating or predicting reality (via retrodiction), and transforming reality through the application of human imagination and knowledge. Action guided by precise planning and preprograming systematically creates or recreates real-world transformations or events. Thought guided by systematic action, observation, reflection, and symbolic representation discovers or invents meanings.

We will now look at man's several language systems, his primary, essential, and most economical instruments for acquiring and applying human understanding. An understanding of languages, symbols, and codes is fundamental for comprehending the interrelationships between the human processes of imagination and inquiry as demonstrated in philosophy and theory and the human methodologies and machine technologies demonstrated in the application of such abstractions to particular physical operations in events and systems.

Symbols originate within a personal cybernetic control system. They are dependent upon personal generation and imaginative application before they can be shared or communicated in either an abstract or social way. Yet the special capacity of human societies for cumulative and incremental knowledge generation requires a systematic and cyclical composition, reconstruction, and communication of personal creations and discoveries. Personal formative experiences must be reconstructed and depersonalized into impersonal facts or interpersonal values in order to become public information. To accomplish this conversion, men need instruments or tools for generating or identifying, fixing or recording, communicating or transmitting, and comparing and evaluating the meanings each individual extracts from his personal experiences.

Human speech and its representations in visual sign and symbol are the most comprehensive and flexible, the most precise and economical, and therefore the most widely used of thought and communications instruments. Cumulative or systematic inquiry cannot progress until words fix objects and events (word referents), and orderly word contexts (statements) fix words in meaningful relationships. The sophisticated inquirer realizes that we have an important interrelationship and inter-dependence among word meanings, word-signs, and things. Statements are meaningful structures of word-signs which order or make sense of the world. Action makes sense of symbolic systems by ordering the world. This is the dualistic representative and regraduative process of systematic inquiry and action.

Prior to the step-by-step application, correlation, and modification of either thought or action there must exist independently man himself and one or more primitive elements in the universe external to man. In addition a human-environmental interaction must occur. There must also exist in meaningful or purposeful human activity the first primitive human idealizations, aspirations, and conceptualizations, together with the rep-resentative signs or symbols which signify, mediate, or simulate these real and ideal world elements. In human inquiry all elements and their significations are open or uncertain in definition and meaning rather than closed or precise.

Incidentally, no matter how perfectly we understand a particular object or event there always remains something unknown, undefined, and uncontrolled about them. But this is not too important. The important human problem is to generate only that use and degree of certainty from inquiry, action, and symbolization which is necessary and sufficient for acquiring appropriate understanding or control.

Machines, on the other hand, as cheerful morons and robots, require that symbolic systems, codes, or programs be perfect (functionally com-plete, linearly ordered, and closed at the beginning and end) before they can be applied or understood. Machines can understand only designative logical-mathematical programs which have unambiguously predefined all elements, operations, and choices. They can use symbols which are as veridically transforming as their mechanical responses. If either the pro-gram or the technology is imperfect, automation breaks down, however. For such language and technology has little flexibility or adaptability. It requires men to periodically update the language program or change the direction of its applications. Automation usually requires critical events or mass production operations to justify the care and concentration necessary to construct precise computer codes and programs.

It is perhaps somewhat unfortunate but completely consistent with man's advancing conceptualizations of reality that systematic human inquiry, as exercised and understood through the application of linguistic theory and method (including semantics, logic, and mathematics) and scientific or experimental method, reflects our recent human awareness that the universe is far more complex than we earlier supposed. However, this very awareness necessitates that we continue to employ and improve our symbolic codes in order to refine, adapt, and simplify through re-organization and eliminative decision-making our orderly thoughts and methodological actions. These adaptively systematic processes must be aided by all of the continuously updated linguistic tools and improved technological machines created by man to augment and accelerate his natural and developed intellectual and physical skills.

Attention now will be directed to a consideration of the functions of man-made signs and symbols as they operate as instruments, messengers, and catalysts in the process of systematic human inquiry and action.

In epistemology, that branch of philosophy which studies human knowledge, there are a number of disciplines or subprograms which have their own unique functions. They are each important in some way for an understanding of scientific inquiry and system processes. These disciplines are all concerned with the systems of signs and symbols which men have developed to record and represent their experience.

Among these knowledge disciplines is linguistics, the broad study or general class representing language; grammar, the study of rules and principles of syntactic structure; and syntax, the study of the arrangement of words in phrases and sentences. And, more important to this context, are the disciplines of semantics, the study of the meanings of word forms, and logic, the study of the principles of formal reasoning and correct inference.

In substance the general system theory of von Bertalanffy and Boulding, and David Ryans' information system theory are essentially synonomous in structure and function with the general semantics theories of Alfred Korzybski, Rapoport, Chase, and Hayakawa, and the theories of psycholinguistics of Charles Osgood and Noam Chomsky. They involve explanations of an empirical and organismic theory of human values, actions and meanings. All of these theories function to put human intellectualism and systematic behavior in their proper place among all other facts of human existence and potential.

These systematic theories and philosophies are semi-open, hypothetical philosophies which are fully universal or eclectic. They can be converted into closed or precise meanings and operations only through an assumptive organization into systems of meaning and value and an arbitrary application of rules of procedure as demonstrated in the properties of reconstructive logic, operations research, and applied technology.

Within an interdisciplinary system context any comprehensive symbolic communications system must contain all of the elements of applied linguistics—objects and events (word referents), words, word relations, word meanings, and word uses. Our combined and applied languages of words and their short-hand symbolizations of logic and mathematics approximate such an ordered structure.

Semantics as a discipline is pertinent to this work because it explains the meaning of words, their semantic content or material reference. Logic as a discipline must be considered also because it explains the structure and order of relationships possible among word-class elements. These two linguistic elements taken together explain many of the systematic functions of verbal language and mathematics as instruments for expediting systematic analysis, purposeful and meaningful human behavior, and machine behavior under human linguistic control.

SEMANTICS—THE STUDY OF THE MEANING OF SIGNS AND SYMBOLS

In the process of human communications through sign and symbol, grammar is concerned with word relationships only. It is not interested in how sentences are related with one another or how words and sentences are related to facts.

Logic is more involved. To the logician meaningful sentences or propositions are assertions of truth statements expressed either in verbal or mathematical terms. The logician is interested in the relations of these assertions to one another (but not to the world of facts). Logical proof is purely a matter of symbolic demonstration of true assertions.

The semanticist pursues another linguistic path. He is concerned with the meaning of words and assertions wherein they can be related to referents (real objects) directly or indirectly. He wants to explain not only the internal logical consistency or validity of propositions but their material truth or fact.

The general semanticist and psycholinguist is concerned with yet a more complex system of relationships. He is involved not only with words, assertions, and their referents in nature but also with their effects upon human behavior. He is concerned with a complete information system of relationships—the chain of external fact to nervous system to language to nervous system to action. (Rapoport, in Thurman, pp. 103–105)

The study of semantics begins with an assumption of the duality of sign or symbol and act or object. The term *semantic* comes from *semanium* (to signify) and means a sign or symbol which represents a material object or event. Semantics is concerned primarily with the material meaning of words.

To be more precise, signs announce, symbols remind. It can be said that animals can respond to signs but only man can use symbols. That is to say, symbols which represent elements that are not immediately present can be used reconstructively and even preconstructively by man.

Symbols direct, organize, record, and communicate meaning with some degree of precision. They take on a mediating determinate quality somewhat independent of either the immediate user or the external world. In other words, they have a reflective and instrumental utility.

The development of the first primitive speech symbols must have originated in a human society which was very active, expressing itself repetitively and ritualistically through ritual, dance, and perhaps cries of terror or joy. (Langer, in Thurman, pp. 3–4) Eventually these verbal noises took a reflective or relational meaning. They became conditional instruments of communication. Later they became informative or evocative elements or codes, at least partially distributed to the respective senders and receivers.

The primitive use of language was certainly characterized by vagueness. Very likely the first words were names or terms used to classify only the most significant objects or actions. Primitive man had few abstract terms that existed apart from material objects. Undoubtedly most of the meaning of early verbal symbols was inefficiently transmitted. The development of mathematical symbolization no doubt experienced a similar and perhaps a later pattern of evolution.

Pragmatic referential is the term used to define general word usage today. The ordinary use of words in daily communication is still far from precise. Any precision necessary in verbal communication is added by shaping the speech context, by the use of contextual clues or signs, or by deliberately establishing boundaries of meaning for particular words— by defining or classifying them.

Words are only partial cues to meaning. Their syntactical and environmental context or event-base as well as their distributed code or history of use (their systematic meaning) partially determine their functional meaning. Thus the meanings of words are conditioned by the past as well as the immediate in the experience of the users. Common meanings of words can be achieved only by common experiences or by commonly assigned or designated meanings.

It is most important that we understand that our common or pragmatic language must be a language with simple yet semi-universal applicability. It must of necessity be flexible and adaptable, universal as well as particular. If words had absolutely fixed meanings, they could be used but once and they could represent only single objects rather than sets or classes of objects. The vocabulary problems of such a usage would be overwhelming.

As our knowledge of the physical universe has increased, we have had to increase our efficient use of symbols to represent it. In order to accomplish this we have had to be extremely flexible in our use of verbal symbols. We use single words or symbols to represent many objects (sets, and even sets of sets) or the same object in many different states. And we use many different words to reclassify single objects, placing them in new contexts or codes of meaning.

It is no accident that the traditional verbal, logical, and mathematical systems of man have undergone a major transformation since the scientific reconceptualization of the universe as a system in dynamic evolution and quantum flux. The history of the evolution of symbolic systems parallels our cumulative understanding of the universe.

This does not mean, however, that a discriminate avoidance of neologisms is not necessary for effective communication. Yet the semanticist is much more concerned with the utility or use of words than he is in their categorical meaning or their level of abstraction.

In analyzing the meaning of words, the use and/or misuse of symbols, the semanticist refers to the purposes of the user. He has found it convenient to consider words as serving one or more of four primary functions.

In signifying referents (objects or events) through verbal symbols, a communicator uses words to designate. (Morris, pp. 95–97) Designators serve the purpose of reporting, informing, and exposing. This use of words provides the receiver of designator symbols with an opportunity for thinking (searching for analytic or reflective meaning). Exposition involves the setting forth of explanations as distinguished from descriptions, narrations, or argumentations.

Appraisers use language to value or express values. This use of words provides the receiver with an opportunity to share a feeling or an opinion.

Prescriptors are words and word contexts used to incite, evoke, or exhort a particular and usually immediate response from the receiver. They are used to motivate toward a specified action.

And we have a fourth mode of signification or use for words—the use of words as formators. The formative or systemic use of language is the use of language to organize, bridge, or translate among the several elemental functions of information giving, valuing, and inciting. Formative usage makes and remakes word forms and structures. It permits the transforming of words in the interest of useful communication between oneself and others. Formative words, word contexts, and word usages permit our rather simple and primitive symbolic system of language to represent our very dynamic and complex universe.

Truth, either significant or consistent truth, in semantic usage is a synonym for the adequacy or appropriateness of signs and symbols. A sign that is truly informative is convincing; it fixes beliefs. A sign that is truly valuative is effective; it communicates values and feelings. A sign that is truly prescriptive is persuasive. And a sign that is truly formative or systemic is correct, precise, or appropriate. (Morris, p. 97)

A designative statement (combination of words and their structural context) that is verifiable by impersonal means such as observation or experience may be considered a public fact. Any other word sentence is either a false statement or an opinion. Designators that denote are the only logically and empirically verifiable word bases upon which to construct logical and verifiable formative statements. Formative or systematic language must be made designative if it is to be considered as factual. Logical discipline is directed toward eliminating or controlling prescriptors and valuators, thus serving this very function. In logical or systematic inquiry we use words and their word structures designatively.

SPECIALIZED USES AND/OR MISUSES OF LANGUAGE

The designative or extensional use of language moves communication from the common or pragmatic level of word usage to the level of objective or semantic referential and explanation. Ultimately the extensional or systematic use of symbols involves the strictly formal or pure level of referential in combination with their content of carefully referenced designative words. Pure mathematics and logic are symbols of pure or universal referential; they are functionally independent of any object reference and require the assignment of such a referent. In general, words or verbal symbols are quite limited in their efficiency for abstracting and for pure communication free from valuative or prescriptive meaning. They are therefore somewhat cumbersome or imprecise in their normal or pragmatic form. Pragmatic language usage is not suitable for systematic analysis. Symbolic logic and mathematical symbols are the most effective symbolic instruments for simulating very complex systems or relationships.

When men use language extensionally, formally, or objectively they seek an exact external referent, a factual base capable of impersonal reference. They literally try to *go out of one's mind* for symbolic terms of referential. They seek objective, public, standardized, and systematized word meanings, word contexts, and word uses. This systematic or scientific use of language employs symbols in their most precise yet in their most widely distributed, independent, or universal manner. Such use of words involves the use of logical or systemic rules which regulate the designative use of words in context.

Another use of general language or pragmatic referential is to communicate connotatively or intensionally. This special type of language function includes the intent of expression using language valuatively, esthetically, or poetically, and the intent of persuasion, using language evocatively, prescriptively, or evaluatively. Connotative language is necessary to communicate simply and rapidly the many value expressions, evocations, and exhortations employed in daily use. Intensional communications generally include an abundant use of adjectives as words of appraisal and exhortation, together with additional significations of emotions. Extensional communications avoid these words and word contexts.

Words used intensionally are offered primarily as subjective creations or opinions, not representations. When men pay more attention to the symbolization of facts or how they feel about facts than to facts themselves, when they indulge in verbal proofs while excluding material reference, and when they seek personal meaning (thinking *inside one's skull*) they are using language intensionally. Intensional language usage also may include the deliberate distortion or transformation of manifest or normative meaning. It serves this gaming function.

The language of intension and connotation demonstrated in the arts and humanities differs markedly from the language of extension and denotation as exemplified in methematics and science. Intension subjectifies; extension objectifies. In the arts and humanities man is not a thing but a drama. Extension and denotation lead to logical conclusions. Intension and connotation lead to imaginative conclusions. Both functions of language must be precise and discriminative in their own particular ways in order to be effective.

A common misuse of language occurs when we do not distinguish between words of designation or extensional reference and words of opinion or intensional reference. In systematic inquiry, when prescriptive symbolization occurs it should be justified upon an adequate base of appraisers which in turn rests upon designators that denote. Designators are necessary first, then appraisers, and only then prescriptors. (Sondel, pp. 161–162) Communications distortion or guillibility occurs when communicators do not use semantic and logical criteria to distinguish between fact and opinion, between the description of the possible or probable and the description of, in many cases, nothing.

Either extensional or intensional language that is misinterpreted is, of course, extremely inefficient or entropic. When we use the expression *It is only a matter of semantics,* we are literally saying that it is only a matter of whether or not we agree concerning what we are talking about or whether our communications have meaning.

The first stage of knowing and communicating (definition) involves establishing a relation between an object or act of reference and its label or name. This stage of definition involves intensional definition. It seeks to explain why a term is applied to an object.

Intensional definitions are usually accomplished ostensively or operationally. *Ostensive definition* is definition by exhibition of the thing to which the term refers.

The meaning of an operational or functional definition is conveyed by exhibiting or demonstrating an operation required to test for the presence of the thing to which the term refers. (Rudner, p. 20) *Operational definitions* explain how a thing works or how it is produced. They are instructions for performing some task, the observation of which is assumed to explain the concept.

Ostensive definitions name objects. Operational definitions name object-relations or events.

The second step in definition is to define a primitive or intensional word tautologically or extensionally, specifying the particular designative meaning of the term that will be used. (Condon, pp. 39–44) In the sentence *The cat is a four-legged animal.* the predicate is tautologous or redundant to the subject. It redefines the term *cat,* sharpening and narrowing the definitional boundaries or classification limits. Predication assigns a term or word to a class. Tautologous or extensional definitions involve fixing agreements about the meaning of words or the system of word usage. They establish agreement about word values but do not insure the quality of reference to material objects. Extensional definition tests the accuracy of a statement against a system of usage but not against an observed reality, or against an unobserved but probable event or explanation (inference), or against one's value system (judgment).

In the complex world of reality man begins his process of establishing meaning by fixing the identity of a single object in the event universe. The object becomes the predicate of its subject (its primitive word or word-referent). It is a specific set or class of elements that is most representative of (always seem to follow) the subjective source.

All consequent stages of objectivity progress from the initial identification of the object, the initial mental representation of the subject (out of mind). Following the first idealization of a material subject is the primitive labeling or naming of the object, assigning semantic content to the object term.

Next follows the classification of the object, relating it to other objects or object-states on the basis of a predicate of statements of relationships of similarity and dissimilarity, more or less, good or bad, etc. This is a matter of establishing a proposition of more or less logical truth or possibility. When enough meaning via diverse systems of classification is established concerning an object, it can be said to be partially known. The final stage in establishing objectivity is the stage of generating a predicate of methodological or material objectivity, the development of a reliable means for systematically maintaining or transforming the object, reaffirming its object classification or transforming it into a new state or class.

DEFINITIONS

Definitions are the predicates or reconstructions of objects and/or their primitive terms. Predicates establish by stipulation or by agreement the meaning of a primitive term or concept.

A primitive term can only be judged as to its meaning or word referent. As it takes on designative meaning or objectivity independent of the user and the immediate context, it becomes more determinate. It begins to represent more precisely the set of attributes and/or events which determine the object.

Designative or extensional definitions signalize the redundant or replaceable (tautologous) in a sentence. They designate the definiens (the term replacing) which is substituted for the definiendum (the term being replaced). Designative definitions often involve a definiendum that is to be displayed in two aspects:

1. Its subsumption under a genus or class, or
2. The characteristics of its differentia (the names and/or functions of its elements).

This stage of definition leads directly to classification and taxonomic classification.

As complicating factors in the task of defining or classifying objects or events, the semanticist explains that term or concept definitions exist on a continuum of clearness or obscurity, of usefulness or uselessness.

Designative definitions name or label observable entities or experiences in the universe.

Dispositional concepts refer to nonobservable or nonmanifest characteristics of observable entities. (Rudner, pp. 20–24) They indicate properties of referents that may be but are not necessarily exhibited. For example, sugar may be defined as soluble or as combustible, but it may never dissolve or burn.

A third level of definitional abstraction involves the theoretical or construct—a nonobservable or nonmanifest characteristic of a nonobservable entity. Theoreticals cannot be directly classified through observation. All metrical concepts, number names, and such terms as time, charge, weight, mass, length, cultural lag, demand, and the names of variables or functors are theoreticals.

Lexical and/or legal definitions are common or normative terms which are frequent sources for designative definition. In some instances psychological or philosophical definitions and theoretical constructs serve as useful analogies for opening up new avenues toward meaning.

In the systematic process of definition, the criteria for determining the definition most appropriate to the situation include the following.

1. The definition should give the essential rather than the accidental characteristics of the element or event being defined.
2. The definition should replace the subject term, advancing it toward more precise and consistent meaning.
3. Definitions should designate observable referents if possible, avoiding obscurity, ambiguity, and figurative language.
4. Definitions can be improved by clarifying, controlling, and developing their context.
5. The most useful definitions are those terms which facilitate experimentally testable reference.
6. Definitions sometimes can be refined by pairing or by the slight modification of a believed unsatisfactory definition.

Definitions involve the making of decisions (judgments) involving the establishing of classes of meaning and non-meaning. They are essentially reducing and clarifying. In the process of definition, objects are assigned boundaries which classify their similarities and differences. Orange is a color (spatial) region between yellow and red. Yesterday is an extended temporal relation to today. Relationships and causal boundaries are asserted and established in operational or functional definitions. These can always be further extended or refined. And they can also be transformed or replaced.

CLASSIFICATION—THE SIMPLIFICATION AND ORDERING OF WORDS

In the process of developing a meaning or understanding of his environment, man has to go beyond the assigning of labels to elements, processes, and theoretical constructs. It is eventually necessary for him to designate definitions which classify, relate, systematize and transform words and word referents. Definition and classification methodologies must eventually progress toward systematization, transformation, and implication.

The semanticist studies words and word referents. The logician analyzes all of the formal relationships that can prevail among individual or class terms. The general semanticist, scientist, and the systems expert are concerned with how the use of words possessing semantic content can be logically ordered and empirically tested in respect to real or simulated human behavior.

The fundamental purpose of classification is to facilitate the grouping of words and/or word referents in some useful and meaningful way. Such grouping serves the purpose of simplifying and ordering the communications system and its referent, the external world. Classification assigns an element or term to a universe of discourse D (for domain) according to some one or more classificatory or categorizing concepts R (for relation or range of meaning). If the classificatory concept or attribute relation (R) of the element applies to only a part of the universe discourse (D), we may divide D into two subsets: D and not-D. (Rudner, pp. 35–36) Thus in any classification there is the ability to ascertain similarities and differences on at least one dimension. Classification is an early step in simplifying and ordering the universe.

In the process of designative definition and classification, man is functioning at an inferential or abstract level of thought. He is making statements about statements. This is sometimes called the second dimension of language, a dimension of word relations that is a concern of the logician as well as the semanticist.

Classification schemes and typologies are called nontheoretic formulations because they imply theorems which admit of logical proof but not empirical testing. They do not contain statements such as hypotheses or theories (sets of hypotheses) which are susceptible to experimental testing. They are outside of systematization which involves establishing a causal relationship and testing an operation, or at least a functional correlation (an induced causal condition which consistently varies or changes with the state of the element or system).

Classification schema, in addition to providing the benefits of simplification, are valuable for grouping words in ways that maximize the recombination and inventive regrouping of data. It is sometimes said that thinking is a matter of employing different tongues (different systems of classification). This type of thinking is based in the process we call analogy.

CLASSIFICATION TAXONOMIES AND TYPOLOGIES

Perhaps the best known and best ordered classification schemes are those of the natural sciences which have evolved into taxonomies. Taxonomies are well-organized and frequently used hierarchial structures of word classes. Such well-established analytic schemata are validated by reference to acceptable criteria of communicability, usefulness and suggestiveness.

Taxonomies are characterized by their hierarchical or part-whole relationship wherein each successively higher taxonomic level or class has at least one new characteristic or condition added. It is more complex and often follows the lower orders or levels of classification in its natural successive development. Taxonomic schemata are usually based upon rather obvious common properties of homologous structure or analogous function or both. Plants and animals are often classified according to physical characteristics although they can be classified by functional characteristics as well.

The taxonomy of educational objectives is based upon a classification schema which assigns its order according to the intended behavior of students, sought by means of instruction and abstracted from educational experience. It attempts to classify objectives which are not as immediately observable or manipulative as most natural science taxonomies.

The utility of classification schema including taxonomies is not directly a matter of making finer and finer discriminations but of discovering only the necessary and sufficient properties or conditions of phenomena which will lead to their eventual systematized and controlled use.

A well-ordered and analytic taxonomic system is one in which all related phenomena can be assigned with little ambiguity. Its hierarchical levels and horizontal categories or classes are both exhaustive and mutually exclusive. In a well-ordered taxonomy each and every horizontal level of defined classes exhausts the universe set, i.e. $a + b + c = u$, and each preceding lower level in the taxonomic hierarchy is a proper subset of the succeeding higher level. In addition, the established classification rules are applied rigorously and consistently. The classification schema should consistently use structural or functional characteristics in defining the elements to be classified.

Schema of classification that are not so well systematized as the better known taxonomies have been used throughout the recorded history of inquiry. Aristotle in his *Organum* proposed that useful categories for classifying elements in the universe might include schema for classifying according to substance, quality, quantity, relation, place, time, position, state, action, and affection. The who, whom, what, why, where, when, and how much schema is another basis for useful classification.

Among the most useful classification categories that appear to exist in system theories are those which separate the operators (defined sources of input action), operations or mediating events, and the operands (the elements being purposefully acted upon and/or the changes in these elements). Subcategories of these major classes can easily be constructed.

In the social sciences, where human understanding is still somewhat primitive, classification according to typologies is found to be an extremely useful device. Typological or polar-opposite conceptualizations such as open-closed systems, urban-rural, *Gemeinschaft-Gesellschaft*, etc. based upon some linking criterion or universe of discourse often lead to a better understanding of the phenomenon under consideration. Positive and negative *idealizations* are common typological schema.

In all classification systems, even in simple typologies, relation or order on at least one scale or dimension is attained. This refines the understanding or meaning of the phenomena being considered to some degree.

There is a logical progression from the nontheoretical schemata of typologies and taxonomies to their theoretical formulation also. In the dyadic relation in which $R =$ father of, where the domain (D) consists of fathers and the co-domain (R) consists of sons and there is some identifiable exemplification of R in D, then D can be divided into subsets and the relation of R to D can be serially ordered. If, in this observed relationship, all of the Rs are serially ordered (are transitive, irreflexive, asymetrical, and connected in relation to D) we can infer a causal relation and assert an imperical condition for testing. (Rudner, p. 36) In such a case, the ordered classificatory relation becomes a hypothesis or a theoretic formulation.

ANALOGY

Analogy is a special inductive or generative process which advances or opens the process of word definition, classification, and systematization. In careful usage, analogy is drawn only between things clearly unlike in kind, form, origin, or appearance wherein there remains some similarity among properties, relations, behavior, etc. Analogy involves establishing word connections in which relations or resemblances are inferred from others which are known or observed. Analogy (a discovered induction) proceeds from the individual or particular to a coordinate individual or particular rather than between natural classes.

Someone once said that although analogy is misleading it is the least misleading thing human beings have to advance their understanding. Analogy is fundamentally an ancient-order or primitive speculative tool which man uses to probe the generally unknown. With it he searches the subjective or *natural* order of existence.

Analogical process is the primary linguistic means for imperfect induction. Its essential uses are four:

In conceptualization (term definition)

In classification (further identification of characteristics via propositions)

In the forming of hypotheses and theories

In the interpretation of what is conceptually remote (implication and inference)

Analogy is useful for several reasons:

All classification is based upon observed or expected affinities (likenesses or differences)

Analogy may assist in creativity which, in turn, permits the generation of new analogies.

Analogy may be employed legitimately in exposition—in the symbolic representation of something relatively difficult and remote from direct experience by means of a more familiar or more easily perceptible situation in direct experience.

For example, we appear to be gaining new insights into the human inquiry process by conceiving it as an analogue of our cybernetic inquiry technologies and machines.

There is some danger in using analogy carelessly, however—i.e., using only logical rationalizations and never resorting to testing. Some sources of this type of rationalizing or speculating error are generated by overgeneralizing on the basis of limited observations. Such imperfect inductions may overgeneralize:

Contiguity—previous time-place connections

Similarity—there are many kinds and degrees of similarity

Emotive congruity—association through vague but perhaps powerful feelings

Frequency—this fourth basis of association tends to produce more useful results (true facts) than the first three

The test of analogy requires one to ascertain whether or not the two elements or systems under consideration are analogous for the purpose for which the analogy is being used. In other words, the test criterion is based upon utility.

Metaphor is a type of figurative speech which uses analogy to stretch the mind, enrich the vocabulary, and perhaps generate invention and imagination. Metaphors, often using such bionic analogies as foot of the table, computer brain, grasping an idea, etc., originate as refreshing ideas which sometimes assume common and legitimate but less colorful or refreshing usage as they are reconstructed or communicated.

In summarizing this unit on semantic meaning, word definition and classification, it seems desirable to repeat that most human communication shifts readily through types of usage, levels of definition, classification, and abstraction. Analogies, metaphors, and other dramatic transformations of language enliven and advance it.

In scientific or systematic language usage, the definition and classification of terms is stabilized. In order to employ language in a consistent and orderly manner the formative and designative communicator will monitor his communications system by asking (Condon, pp. 43–44):

> Is the term, class, or statement tentative or final?
> Was the statement obtained inductively?
> Is the statement absolute or probable?
> Does the abstraction have a specific use?
> Does the abstraction have a systematic, universal or continuing use?
> Does the formulation move from a descriptive function toward a predictive-explanatory function?

Such precise determinations of semantic meaning plus subsequent systemation via logical and scientific testing are essential for systematically representing material systems through symbols. The reader is reminded, however, that logical and scientific methodology are mid-range techniques or methodologies. Their combination of a limited amount of semantic content (representative of the homogenous in certain particulars) with the limited mathematics or logic (representative of universals) is always only a partial (probable) explanation. It is inherently and by intent more effective at explaining process or act meaning than it is in explaining product, value, or action meaning.

4

Logic—The System in Symbolic Systems

Concerning the significance of systematic inquiry John Dewey wrote:

> Inquiry is the controlled or directed transformation of an indeterminate situation into one that is so determinate in its constituent distinctions and relations as to convert the elements of the original situation into a unified whole.
>
> (Dewey, *On Experience, Nature and Freedom*, p. 116)

This is a capsule summary of the process of system analysis and synthesis. In order to accomplish this transformative process, man has had to develop symbolic systems capable of unambiguously representing or reflecting reality in past-to-present and present-to-future transformations. Logic is the science of such symbolic systems.

Logic is a kind of linguistics which seeks to develop rules or specifications of permissible combinations and permutations in symbolic systems which are capable of representing elements and operations in real systems. (Rudner, p. 12) A symbolic system becomes deterministic relative to some specified set of characteristics only if there is a deterministic theory or set of rules that predicts or explains its states in relation to those characteristics.

A being whose intellect is infinitely powerful to permit perfect induction, explanation, or interpretation of his universe of states and transformations would have no interest in logic or mathematics. For he could immediately understand everything he could experience, perceive, and define.

However, any manageable symbolic system of such completeness and consistency would have to be infinitely complex yet elegantly simple. It would have to be similar to the combination of a purely formal, abstract, and uninterpreted axiomatic system or calculus that logic now is and a designative semantic system of classification and enumeration.

Abstract Boolean algebra exemplifies for the system expert the essence of present calculi. It is a powerful symbolic logic or calculus of sets or classes and syntactical or operational rules; but it has no meaning in the literal or semantic sense. Its elements are unassigned universals. In order to add meaning or possibility of interpretation to this logic, it must be given particular semantic content (words and word referents). It must be applied to material reality. All algebraic variable symbols or forms must be given particular values.

Pure logic, including its extension into pure mathematics, involves pure abstraction. It is thus applicable to everything and yet representative of nothing in the material universe. Bertrand Russell expressed the significance of pure symbolic form and structure in this manner: "Mathematics may be defined as the subject in which we never know what we are talking about, nor whether what we are saying is true."

Scientists understand the essence of pure form and structure. They know that they do not directly prove theories or hypotheses. They only prove instances of them, particulars which are systematically assigned semantic referents which are essential to their meaning.

Logic may be called the grammar of reason. It is the science which establishes norms or rules which distinguish sound thinking from unsound thinking. Logical procedure involves progressive systematization and generalization, beginning with first level symbolic abstractions (primitive terms) and progressing to designated class terms and their permissible, implied or inferred relations.

Although logic is a science, reasoning is an art. Logic may be taught but reasoning requires exercise of the imagination, a sense for the relevant and the persistent. Alfred Ayer, noted logician, wrote that our intellects are unequal to the task of abstract processes of reasoning (inference and implication) without the assistance of intuition. (Ayer, pp. 33–34) Inference remains primarily a human creative operation even when implication is dependent upon reference to external relation-rules of the system.

Logic proceeds from truth (material existence or significance) to consistency or system. It seeks to distinguish the consistent from the inconsistent. Any formal symbolic or linguistic procedure which is self-consistent will satisfy the requirements of logic.

Logic and deductive-analytic systems are never concerned with existence except in a hypothetical sense. Their proofs are established through formal symbolic demonstrations. Only applied logic and mathematics are of direct interest to most natural and social scientists, including educators and administrators. These latter groups are interested in logical structure only when it is combined with or is applied to the semantic content of their particular problems or concerns. They are interested in material truths and/or probabilities rather than logical truths.

Strictly speaking, logic is a deductive-analytic science, progressing from well-formulated (near-perfect) system to unique event application. In this sense it is primarily reconstructive or informative. It functions as a deductive-analytic, closed, or categorical system; it serves an interpretational or explanatory function when so used. Abraham Kaplan calls this strict logic *reconstructed logic*—a reconstruction based upon retrodictive, recursive, or past-to-present experiences of the system. (Kaplan, pp. 3–11) Thus the process of perfect or systematic enumeration, explanation or interpretation, progresses from system to event. It is deductive-analytic.

Kaplan also defines a logic-related process which he calls *logic in use*. This is a process of induction. It functions to generate concepts, classes or sets, taxonomies, paradigms, models, hypotheses, and theories. It involves the exercise of inductive inference.

INDUCTIVE INFERENCE: LOGIC IN USE

Induction means causing, producing, initiating, or generating. Inductive inference involves the process of moving from that individual reality which is unknown or ill-defined to symbols which assert identity and relationship classifications, including cause and effect. Induction expands or opens knowledge and reduces certainty. Deduction increases certainty at the constriction of knowledge.

As stated in Part One of this book, inductive method in logic and science invents symbolic or material elements. It generates symbols which explain (transform) the old in the light of the unique. Induction progresses from the particular to the particular or from the particular to the general; deduction progresses from the general to the general or from the general to the particular. Inductive reasoning concerns generating means of attaining reliable beliefs by checking the reliability of statements through application.

Inductions—imperfect, heuristic, or uncertain—are essentially open, tentative, and hypothetical. With one or two tests they are assumed to be valid. Only one contrary instance is needed to negate the truth of such generalizations, however. But great care must be exercised to make sure that what appears to be a contrary instance is so in fact.

The general principles or laws of logic were derived through inductive inference (hypothesis-making) and confirmed via deduction (hypothesis-testing). The fundamental logical principles of implication employed in the hypothetical syllogism (If this is true, then that is true) is an example of an induction. The general laws of logic obtained via induction include (Cohen and Nagel pp. 183–185):

1. The law of identity—whatever is, is $(a = a)$.
2. The law of contradiction—nothing can both be and not be $(a - a = 0)$.
3. The law of the excluded middle—everything must be or not be $(a + -a = 1)$.

All of these universals (logical laws) of reasoning are hypothetical. They tell us that if one thing exists another must exist, or more generally, if one proposition is true, another must be true.

It is only by means of inference that logical analysis and induction go beyond the statement of tautologies or the report of observable facts. Inference involves intuiting or creating hypotheses about relationships between the unknown and the known. Skilled logicians discover their implications (necessary connections among stated logical elements), but they must create their inferences.

The primary purposes of inference (reasoning from the known, observable, and experienced to the unknown, unobservable, and nonexperienced) are to assert and discover causes, to facilitate predictions, and to improve descriptions. We can justify inductive inference only by its generative or transformational utility. Inferential utility can be methodologically controlled by evaluating or analyzing inductions by these procedures:

1. The method of differences—only one variable is changed.
2. The method of concomitant variation—noting correlation: if a is > (greater than), then b is > (greater than).
3. The method of agreement—noting critical incidences.
4. Joint method of agreement and difference—both variability and critical incidence are noted.
5. Method of residues—all alternatives are exhausted.
6. The methodology of causation—causation must precede effect in the operation.

In the method of inductive inference or discovery, hypothesis-making involves asking and tentatively answering the philosophical question of "Why?" which implies "How come?" In argumentation (a synonym for the induction on an hypothesis) there is a conditional premise to the effect that if one thing is the case, then a second thing is the case. There is another premise to the effect that the first thing is the case. Finally, there is a conclusion to the effect that the second thing is the case. This is indicated symbolically as: if p, then q $(p \to q)$; p, therefore q.

Induction exercises intuition, insight, imagination, and risk-taking as well as analysis. It is essentially unpredictable as it involves some degree of uniqueness and particularity. No one has demonstrated that there is or can be a systematic logic of induction or discovery.

The inductive inquiry style or set for hypothesis-making or productive heuristic behavior allows actuation of the imagination, the entire memory repertoire. This is encouraged by evaluative-free expression, divergent thinking, and the generation of alternatives to the normative, predictable, or general conditions—going beyond the known. Induction tends to work best (display the greatest immediate utility) when its sampled instances are representative—that is, if the trials are based upon some prior knowledge or experience and if the material is homogeneous (relatively well-known and consistent). However, in induction the conclusions presented contain information not present even implicitly in the premises.

Typically, useful propositions or hypotheses originate from within the personal experience of an inquirer. Analogy (reclassification) is useful in hypothesis-making. Identifying and projecting better resultant behaviors, using positive or negative typologies, and model building sometimes prove fruitful in induction.

In establishing a useful proposition or hypothesis, an investigator might ask himself:

1. If the objectives of the program or problem solution are to be realized, what would be the observable behavior (evidence)?
2. Where and in what situations would this evidence be observable?
3. How would the evidence be collected and organized?
4. How would the evidence be appraised and its significance determined?

DEDUCTIVE ANALYSIS AND INFERENCE

The system of deductive-analytic logic presupposes primitive terms, word referents, and their connectiveness. Thus we should keep in mind that all meaning or knowledge has something which is asserted a priori and something which is relative, both of which do not properly exist. This set of presuppositions enables meaning to be at least partially abstract, universal, and independent of the object(s) known and of the knower. And it has possibility or potential prior to being known.

Meaning becomes synthesized, correlated, or fixed only through the inquiry act and the perception of the knower. Complete explanatory knowledge (knowledge about, know-how, and knowledge of) is achieved only a posteriori to systematic inquiry. Knowledge about (awareness of existences or hypothetical meaning), and know-how or evidential meaning may precede and accompany experience, but knowledge of (determinate meaning or complete explanation) and wisdom (judging the value of knowledge) must be a posteriori to inquiry.

A priori truths or statements of inductive inference can be systematically validated only by logical analysis plus empirical representation and replication (hypothesis-testing). This is the particular and unique function of logical deduction and the scientific method.

It is quite important to the systematic inquirer that he becomes convinced that there are no intrinsically undemonstrable logical statements or propositions. No assertions of truths can be maintained that are not open to both logical analysis and empirical proof.

Human valuation or incitive expressions per se are not logical propositions but only pseudo-propositions. They have no independent, consistent, or instrumental value or meaning unless they are at least partially transformed. They remain outside of orderly validation and verification.

Once concrete or definitive statements or hypotheses are generated, the chief business of the canons or laws of logic is to delineate correctly the functions of one or more relations or conditions which hold between any set of premises (prior semi-independent statements or propositions) and a conclusion or consequence. Deductive or systematic logical analysis involves the establishing of categories of elements and relations in such a well-defined or closed system. Deduction or systematic presupposition always implies that a large segment of the field or discipline is formalized, structured, or known. Extensive enumeration of objects and relations or events is possible. (Cohen and Nagel, pp. 30–77)

The basic units of traditional deductive logic include:

The term—which designates the class or set, the smallest unit of logical reasoning;

The proposition—a statement of an hypothesized judgment about some term or terms;

The syllogism—the organic unit of logic. It entails a relation among three propositions. The first two are premises, causes, or evidence of the third, the conclusion or consequence.

Deductive inference is the system for establishing by means of inference and implication the consistency among terms, propositions, and syllogisms. It establishes consistencies from truths. It expedites the evaluation or judgment of these logical elements.

The Term

Concerning the systematic use of terms which designate, we induce objectivity into terms by their conceptualization as predicates of ideas, tautologous refinements and extensions of word referents. Terms are judged as to their meaning. They exist in a continuum of clearness or obscurity. Designation which involves word referents and word agreement explicitly clarifies the meaning of a term.

The Proposition

Propositions may be defined as statements (sentences) or arguments which describe an object or object-referent (term). They include all statements which can be said to be true or false, consistent, and independent of the observer.

All propositions are statements which assert or deny something of something else. That which is asserted is the subject and that which is asserted about the subject is the predicate.

Statements of asserted propositional relationships transform qualities into more or less determinate distinctions or relations. This makes the quality into an object or event capable of rational consideration, eventual testing, and establishing as fact. Anything that cannot be stated and described consistently remains subjective or ill-defined and outside of meaningful consideration.

A demonstration of several types of statements might facilitate a better understanding of the meaning of logical propositions and their constituent properties and conditions. Let us consider the following statements:

1. The cat is a four-legged animal.
2. The cat is meowing.
3. The cat is ill.
4. The cat is the finest of pets.

The first of these sentences is a tautology, an agreement about the meaning of the words employed and the system of usage. It agrees on value-terms but not on observations. (Condon, pp. 73–83)

The second sentence is a description. It is immediately open to confirmation by those capable of observation. It has ready object reference. It is easy to establish as fact by reference to existence via observation or memory. Inference is unnecessary here. This sentence permits a ready agreement on both terms and observational referents.

The third sentence is an inference. It requires some uncertainty or guessing about the unknown on the basis of the known. Its confirmation ranges from immediately verifiable to unverifiable. It is only probable.

The fourth sentence is a value judgment. Agreement of terms and relations here is of little importance. Valuative or prescriptive statements cannot be easily or consistently verified.

These statements explain the several kinds of subject-predicate relations possible in a proposition. Tautology is a matter of agreed definitions. Statements of fact involve ready confirmation through perception and memory. Asserted truth or falsity beyond the immediate requires necessary inference, the relation of unknowns to knowns, and some propositions assert value judgments or preferences wherein the meaning of terms and observational referents may be personal rather than consistent or systematic. It is necessary for the systematic inquirer to recognize that facts are not inferences and that inferences are not value preferences. However, skilled logicians through the expert use of inference, implication and tautology can develop hypotheses assumedly representative of value judgments and inferences that can be tested experimentally and proved indirectly.

We cannot and need not infer anything about object terms or factual events or statements. Only when we reach the point of determining unobserved relations among terms are we involved in inference.

A strict definition of a proposition states that it is a class of sentences which has the same significance for anyone who understands it. Logicians require that every proposition which we can understand must be composed wholly of constituents with which we are acquainted. Deductive reasoning involves drawing out the implicit beliefs which must be assumed if certain other beliefs are accepted. It determines if it is possible for the conclusions to be true without the premises being true.

A sentence has literal meaning if and only if the proposition it expresses is either analytic or empirically verifiable. It is analytic when its validity depends solely upon the definitions of the symbols it contains. This condition or relationship is called formal, tautologous, or strict implication. Formal propositions which cannot be denied without self-contradiction are analytic.

Material implication requires the assertion of truth and testing for it empirically. However, no proposition itself can be demonstrated by experimental methods alone. Material implication also requires establishing formal implication and tautology. A proposition is proved logically when and only when a premise implies that proposition and when the premise itself is true. Analysis only establishes formal implication, however—a necessary connection or entanglement between the subject and predicate.

In logical verification we thus find two questions involved:

1. Are the propositions offered as evidence true?
2. Are the conclusions so related to the premises that the former necessarily follow and may thus be properly deduced from the latter?

The first question involves the experimental, applied, or material issue. Logicians are concerned with formal implication—the second question.

In the process of determining useful premises upon which to build inferences, logicians consider these sources:

1. Statements of fact—verifiable by observation or experience and the most reliable source of premises.
2. Judgments—conclusions inferred from facts.
3. Expert testimony—permitting inferences from authoritative statement; the source of the statement may be either fact or judgment.

It is only by means of inference that men can go beyond the mere statement of tautologies or the report of observed facts in their thinking. Formal logical analysis involves inference from one proposition to another or to a class of propositions by a pattern of implications until a test can be made of the negation of the proposition without self-contradiction. Skilled logicians expertly create chains of inferences. By exercising their rules of logic, they discover (test for) the necessary logical implications.

The Syllogism

Logical inference, to be valid and useful, requires the establishment of implication between propositions. Thus we arrive at a consideration of the most significant or classical form of deduction—the syllogism. In the syllogism we reason that if two things are known from premises, a third thing can be concluded, because the third thing is implied (is a tautological or alternate way of restating the information already given). The syllogism is a mediate inference. It is an inference by elimination of one or more terms in the premises. For example:

> All men are mortal,
> Socrates is a man.
> therefore
> Socrates is mortal.

Within this process of mediate inference or implication, logicians have established a system of terminology. The term which is contained in both premises is the middle term. The predicate of the conclusion is the major term and the subject of the conclusion is the minor term. The premise which contains the major term is the major premise and the premise containing the minor term is the minor premise. The order of statement, therefore, does not determine which is the major premise. In the process of logic there are a number of axions accepted as true (useful) in deductive inference. The following logical axioms of quantity and quality prevail in analyzing the syllogism (Cohen and Nagel, p. 79):

> Axioms of Quantity:
> 1. The middle term must be distributed at least once.
> 2. No term may be distributed in the conclusion which is not distributed in the premises.
>
> Axioms of Quality:
> 3. If both premises are negative, there is no conclusion.
> 4. If one premise is negative, the conclusion must be negative.
> 5. If neither premise is negative, the conclusion must be affirmative.

These axioms, together with the principles of inference, are sufficient to develop the entire theory of the syllogism.

Within the system of logic there are four primary types of syllogisms. These are:

1. The Categorical Syllogism
   MP All a is b
   mp c is a
   C c is b

2. The Conditional or Hypothetical Syllogism
   MP If a, then b
       (If a given condition exists, a second will happen whenever the first is satisfied)
   mp a
   C b

3. The Alternative or Nondisjunctive Syllogism
   MP Either a or b
       (If one is true, the other may or may not be true. However, in an alternative syllogism either/or does not imply that the altenatives are mutually exclusive.)
   mp Not a (or not b)
   C b (or a)

4. The Disjunctive Syllogism
    MP Not both *a* and *b*
        (Mutually exclusive statements. If one is true the other must
        be false. The disjunctive syllogism permits a valid conclu-
        sion only when one of the two statements is affirmed in the
        minor premise.)
    mp *a* (or *b*)
    C   Not *b* (or not *a*)

The reader will note the varying properties of quality and quantity upon which the propositions in the four types of syllogisms are based. Propositions are universal, definite, particular, and/or indefinite. This reflects the quantitative dimension of the proposition.

If we think of a proposition as a relation between sets of individual elements, an affirmative proposition asserts the inclusion of one class or particular in another. An affirmative proposition is true if and only if something makes it so. It requires proof. A negative proposition infers the exclusion between classes or sets and particulars. It is true only if nothing makes it false (the null hypothesis). The qualitative dimension of a proposition is determined by its affirmative or negative characterisics.

In a universal proposition the subject term is always distributed; in a particular proposition the subject is undistributed. Affirmative proposi-tions do not distribute their predicates. However, the predicate is dis-tributed in a negative proposition.

In a systematic inquiry, universal propositions always function as hypotheses. They must always be possible; and they must always be general, dealing only with relations among sets or classes. Thus scientific or hypothetical propositions assert that some object is the member of a class or set or that two object-sets have some relation (correlation). In regard to relations, two propositions are said to be dependent if they exhibit some relation or correlation.

The principles of syllogistic reasoning are based upon the laws of identity, contradiction, and the excluded middle which were presented earlier. These original axioms have been extended in modern logic and mathematics to include the principles of commutation, association, dis-tribution, tautology, absorption, simplification, composition, and the syllogism, all fundamental principles of Boolean algebra, set theory, and symbolic logic.

In careful systematic analysis we follow this deductive procedure. If we have a set of *n* propositions and if one of them is a logical conse-quence of the remaining *n-1* set, that one will be redundant (tautological) with the remainder and may be discarded from the set. When no further reduction of this sort is possible the propositions remaining in the set are said to be independent. They then can be more objectively identified, enumerated, and interpreted.

Of the four forms of the syllogism, the hypothetical syllogism or hypothesis is the classical form of greatest use in scientific or systematic inquiry. A hypothesis always asserts a conditional, however. It implies a *can do* rather than a *does do*. Thus it can be altered or reversed (arbitrarily changed from true to false). It asserts that only if the antecedents are true, the consequence must also be true. In establishing logical implication and scientific application, a hypothesis must be capable of being confirmed as valid or invalid and it must assert something about the real world.

A function, the constituent or determinate predicate relation(s) of an element and an event, is the result of a combination of conditions. Functions may be regarded as syllogistic statements. They are a self-persistent set of conditions. Conditions may be represented logically as propositions. Conditions can be replaced by (can be analyzed into) two or more properties. Properties are independent terms designating sets or elements. Thus we can apply the analytic laws of logic to natural referents and assert the probable outcome of functions or operations in a system.

THE GENERATION AND FUNCTION OF HYPOTHESES AND THEORIES

The final explanatory argument or proof to be derived from systematic methodologies effects a transition from logical formulations or calculi to their use. This type of explanation differs from prediction primarily in relation to the pragmatic vantage point of the inquirer. Material systems actualize logical proofs. Scientific method and system test hypotheses and theories by inducing specific material particulars (events) into a more extensive environment (system). This essentially generates or originates a new empirical theory or possibility.

Explanation is retrodictive to system; it follows one or more events in a system (Rudner, pp. 63–67) Prediction concerns the generation of a symbolic or simulated explanandum, hypothesis, or theory, antecedent in space-time order to its systematization.

Scientific hypotheses or theories come into being by deductive or discovery means rather than by inductive or creative means. The scientist foresees or discovers the possibility of a law or generalizable-systematic statement before he is in possession of the evidence which justifies its creation. The only inductive element in theory is creation which emphasizes explanation after the fact—the testing, application, or materialization of the theory. Inductive theorists are activists and risk-takers. But they have difficulty in generating or extending explanations of extremely complex phenomena. For this reason symbolic speculation, reclassification, analogy, hypotheses-making, model and paradigm construction, and theory building are gaining in importance as men understand more and more of the characteristics of nature.

Hypothetical inference may be called reasoning retrodictively or recursively, from the consequent or solution to the antecedent. (Peirce, pp. 45–46) It is often based upon a self-induced but definitive representation of a value or goal to be reached. Yet, hypotheses or theories are not value statements but inferences of a more substantive or objective nature. At best the test of a hypothesis or theory explains what is, or can be, not what ought to be.

A theory is a symbolic hypothesis or set of hypotheses which anticipates what is. Theory becomes useful only when it serves as the explanandum of a full-fledged scientific prediction (when it is deductive-analytic).

Theories are statements about concepts or variables which build a structure explaining the relations among them and their factual referents. They are supposed to be compatible with existence, tolerating no exception. Theories are judged by their truthfulness, precision, and consistency or comprehensiveness. Theories and facts are reciprocally interdependent for their meaning and their order can be reversed.

Models and paradigms are partial theories that are analogies of relationships among word referents which tolerate some facts that are not consistent. They are judged by their utility rather than their consistency. Models or paradigms may be physical drawings or replicas, or they may be abstract symbolizations. They may be mathematical equations, verbal statements, symbolic descriptions, graphic representations, electro-mechanical or iconic devices. Statements of physical laws are often useful models or analogues that serve as partial theories in the explanation or prediction of human behavior. Occasionally two contradictory models may be used simultaneously to interpret some phenomenon such as the wave and photon models for light.

The association of a calculus or method of numbering and classifying objects and operations in a system is not a genetic, chronological, or practical association, but a logical one. Any theory building requires the introduction of new concepts (analogies or reclassifications) which, although introduced by means of old definitions, replace rather than increment the elements in the theoretical set.

The functions of theories or hypotheses are to:

1. Assist in observation, testing and interpretation of systems.

2. Assist in classifying phenomena.
3. Assist in formulating constructs, abstract conceptualizations of intervening variables or unobservable phenomena (dispositionals and theoreticals).
4. Assist in summarizing, predicting, explaining data not directly available to observation or memory.

Computer programs are precise and detailed materially implicative model or paradigm statements which unambiguously enumerate or simulate a transforming or operating system.

Theories, hypotheses, or axioms are studied for the propositions (set of assumptions) they imply. Fertility in symbolic explanation and/or material application are the criteria of their value.

According to James Jenkins (in De Cecco et al., p. 310), in any fully developed scientific theory there are three distinguishable levels of linguistic and scientific relationships. The first level involves the direct observation or immediate apprehension of raw sense data or its observed effects upon scientific instruments. All sciences involve this level of experience.

The second level of scientific theory development involves addition of linguistic concepts and the laws or rules which consistently summarize or define all hypothesis-predicting relationships not yet observed. These concepts and laws, to become fully meaningful, must be unambiguously related directly or indirectly to Level One events. They function as operational definitions. Any system which uses concepts which cannot be operationally defined is ambiguous or prescientific at best and possibly meaningless, doomed to remain always a priori to verification.

The third level of scientific theory development involves the level of formal logical and mathematical concepts. It states the rules of acceptable logical inference. The interpretation of this level or system requires placing its elements and relations in correspondence with Level Two concepts which, in turn, are consistent summaries of observed or predicted Level One events. Scientific theory must relate sense data, semantic meaning, and logical inference.

It is in the nature of theories that they must tend toward producing this integration of meaning by creating discrete categories for classifying independent theoretical entities. If a set of axioms or theories is independent, it is possible in that system to make a sharp distinction between assumptions of a priori truths (premises) and theorems (their implied conclusions). This property of independence is most important. A set of propositions becomes categorically independent if it is impossible to deduce any one of the remaining theoretical concepts from the others.

The dualistic and regraduative methodology of the scientific-logical discipline or method is not a matter of its transient or heuristic techniques, however. It is primarily a matter of the logic or system of justification. All science is characterized by (is dependent upon) this common logic of justification by which it accepts or rejects hypotheses or theories.

Substantive or semantic ignorance is not a fault of science or system per se. It is an induced a priori condition involving incomplete experience and information distribution. The deadly scientific or systematic sin is the sin of methodological ignorance—the failure to apply the rules of logical analysis and the rules of the experimental and system methods.

5

Symbolic Logic and Mathematics: Systems for Ordering or Scaling Meaning

Any language or linguistics which seeks to efficiently represent material phenomena must exhibit four properties. (Rudner, p. 89) It must:

categorize and classify—assign terms to objects;

permit abstraction—allow the symbols to represent the phenomena universally (independent of the presence of the phenomena);

establish a form or structure of relations among classes (rules of transformation);

provide symbols which are complete; they can indicate both the objects (classes and sets) and the permissible transformations or operations in the system.

An adequate symbolism indicates what is constant and what is variable. And it must do this with precision and brevity.

Modern symbolic logic and Boole-Schroder algebra with its revisions and extensions into modern mathematical set theory are a very effective linguistics or calculus for simulating complex systems of elements and relations. Boolean algebra permits the expression of highly complex tautologies or equations in conveniently simply form (1 for true and 0 for false) for visual display and manipulation. It is adaptable to binary mathematical notation although it is not identical with it. Advances in human inquiry have often been accompanied by such significant revisions of symbolic notation.

As a result of scientific theory development and experience gained through the application and testing of the new logical systems, logicians and mathematicians have concluded that the methods of mathematics are applicable not only to the study of quantities but also to the study of any ordered realm or system whatsoever. Dewey and Einstein were among the great modern philosophers and logicians who predicted the ultimate triumph of the statistical method.

Symbolic logic or Boolean algebra is a logistic or deductive-analytic symbolization that establishes terse symbols which designate terms in extension and which symbolize the laws of relations among these terms. It asserts that the calculus of propositions may be derived from the calculus of classes and that propositional functions or arguments (propositions whose truth value is only probable) and, finally, systematic operations can be represented in a similar manner.

Perhaps the simplest manner by which to introduce logical-mathematical analysis and symbolization can be effected by presenting it in the order normally assigned to traditional mathematical scaling (determined by the precision and/or detail of the scale).

Prior to any ordering or scaling of meaning or judgment of logical-mathematical statements there must exist elements or actions and their primitive word referents. The relations between these primitive objects and events are relatively unknown; they involve near-zero meaning, blind trial and error control, and chance. Any units facilitating the identification of objective properties in this event universe are absent. Scaling of such units of meaning provides consistent measures no smaller than the entire universe being assessed.

THE NOMINAL-OBJECT SCALE OF MEANING

The first step advancing inquiry toward ascertaining greater meaning of an event or system is accomplished by assigning names to specific elements or actions in the event universe. Definition of objects involves selection and abstraction, the first extraction of elements or subsystems smaller than the whole of the event universe or system. It involves the reconstruction or fixing of the objects or actions in symbolic form also. It organizes and/or habituates the relation between the object and its representative image. Mutual distribution or sharing of the concept and its meaning are now possible to a greater degree.

Within the nominal scaling of meaning and judgment, the original distinguishing of specified objects and events is accomplished, but there is no ordering of the relationships between a designative term and any other word elements in the descriptive universe. However, the fixing of the relationship between the object and its symbolization involves the transformation of a particular object to a universal set or class term. At least one set (defined as any well-defined collection of objects, numbers, people, events, or anything) is defined.

THE ORDINAL SCALE OF MEANING AND JUDGMENT

The fundamental property of the ordinal scale of meaning, judgment, or measurement is the property of relation, classification, or comparison —the property of association between identifiable external objects and referent set terms. This ability to compare, relate, or associate permits humans to establish a rank ordering of classification and/or preference between any two objects or events. It permits description to include reference to similarities and differences and reference to preferred or ordered values. As Charles Peirce said: "Good and bad are feelings which first arise as predicates." (Peirce, p. 33) Ordinal scaling and judgment permit the statement of propositions which can be assessed as to their truth.

RELATIONS

The concept of order or relation is fundamental to logic. The awareness of relations between object-classes or sets permits making explicit any statement about an object in reference to another object. These relations may also include part-whole, particular-universal, subsystem-system or hierarchical relationships. Being a member or element of a set is a relation.

Relations are essentially dyadic in nature. They establish connections, interdependencies, and reference. A relation may be defined as a set of ordered pairs, a rule of correspondence.

In dyadic relations the first element of the set of ordered pairs is called the domain (D). The second element of the set is called the relation range (R). The relation or rule of correspondence (R) assigns to the domain some one member of the range. It involves an association of the variables (set terms) a and b which pairs each admissible a with one or more values of b. The degree of connexity between ordered pairs is related to whether the relation holds between every pair of the set collections. (Rudner, p. 36)

The primary properties of logical set relations include:

1. The property of reflexiveness (the principle of identity)—a relation holds between a term and itself $(a = a)$.
2. The principle of logical contradiction— $a \neq a$ $(a = -a$ [*not a*]) is a logical contradiction to possibility. This may be restated as $(a - a = 0)$
3. The law of the excluded middle asserts that something must be either a or $-a$. It must exist or not exist. The Boolean equation for this is $a + (-a) = 1$.

The most significant dyadic relations that can be logically inferred between two objects or sets are the relations referring to degrees of symmetry and transivity. Specifically these types of dyadic relations include relations that are (Cohen and Nagel, pp. 113–115):

> *symmetrical*—equal to, unequal to, near to, distant from;
>
> *asymmetrical*—if, when *a* bears a relationship to *b*, *b* cannot bear it to *a* (larger than, smaller than, etc.);
>
> *nonsymmetrical*—if, when *a* bears a relationship to *b*, it is logically indeterminate whether *b* bears it to *a* (Jack loves Jill);
>
> *transitive*—if, when *a* bears a relationship to *b* and *b* bears that same relationship to *c*, it necessarily follows that *a* bears that relationship to *c* (aRb) (bRc) = aRc (richer than, porrer than);
>
> *intransitive*—if, when *a* is so related to *b* and *b* is so related to *c*, it is impossible for *a* to be so related to *c* (mother of);
>
> *nontransitive*—if, when *a* is so related to *b* and *b* is so related to *c*, it is logically indeterminate whether or not *a* is so related to *c* (if Bob is friendly to Bill and Bill is friendly to Joe, nothing about Bob and Joe can be implied).

Symmetrical relationships are reciprocal or reversible, permitting alternation of the order of the pairs. The inclusive either-or logical relationship is symmetrical.

Nonsymmetrical relationships may be reversible; this possibility or probability is uncertain.

Asymmetrical relationships are necessary to establish serial order that is irreversible.

Transitivity involves the consistency of relations among three sets or set terms. Transitive relations carry across ($a > b > c = a > c$) where $> =$ greater than.

Intransitive relations serve to determine impossibility (aRb) (bRc) $\neq$ (aRc) where the relationship is *father of*.

Nontransitive relationships limit any possible implications involving the transitivity of relations among three elements.

Among the six types of relationships just discussed, the properties of symmetry and transitivity are semi-independent of one another. Thus we may obtain nine combinations of symmetrical-transitive relations. Of particular importance to persons applying rules of logic to systems is the asymmetrical-transitive relationship cambination. It establishes serial order among the members of three or more sets. This pattern of relationships also establishes betweenness. In short, if a term has a successor, it has an immediate successor, and, if it has a predecessor, it has an immediate predecessor. The reader is cautioned to remember that betweenness does not alone determine direction or serial order, however.

The asymmetrical-transitive relationship law also establishes the possibility of the hypothesis.

Thus we come to a particular and very important kind of logical relation—the relation of function.

FUNCTIONAL RELATIONS

A fundamental problem of logic and mathematics is to determine whether or not a relation is in fact a function.

Function determines dependence, predication, and order among two or more elements of a set. It is initially ascertained by a discovery of correlation—the determination that one thing reacts or varies consistently and dependently with another.

A function is a single-valued or ordered relation in a set of ordered pairs (x, y) where each x is a member of a set (D for domain) and to each x there corresponds a unique y. The function is categorical, complete, consistent, or necessary if the isolated relation can be ascertained as singularly causal in respect to a specified effect—that $fx \to y$ (the function of x implies y) where x is the domain of f (function) and y is the co-domain of f. In such a closed situation, where the relation is transitive, irreflexive, asymmetrical, and connected, causality can be inferred. Such a logically ordered relation represents a determinate system, one which, when actualized, produces exactly as expected or desired. Scientists like to develop precise logical or mathematical statements (equations or laws) which model or predict a function or series of functions.

The range of a function includes all those elements in R which appear as the image of at least one element of the domain. Specific terms within the range of meaning of a variable in a function are called the values of the variable. The rest of the statement is the invarient part of the function. In logical or mathematical proof, truth depends solely upon the consistent relationship of the conjunction or addition of variables. It takes but one contrary example (disjunction) to disprove a logical statement.

In ascertaining a single functional relation, where $fx = y$ and where f is a necessary and sufficient condition for y, the function is operationally equivalent to its co-domain or effect. To repeat y in the equation is an unnecessary tautology. When we define an object operationally, we are in effect replacing the dependent or output effect or condition with its equivalent functional, operational, or mediating condition.

Functional correlation often seeks to ascertain the possibility of isomorphism, wherein the unique relation established becomes discrete and additive. In such a situation all properties become cumulative, associative, and incremental. Equalities added maintain equalities. This is not generally accomplished in complex situations. Functions may be equal, constant, identical, isomorphic, a product, or an inverse.

Negation or subtraction is the null function of addition.

Universality and particularity, properties of number, are also functions.

In dyadic or triadic functional relations or correlations, functions may vary in number as well as order. There may be: many-many correlations (x is a creditor of y where both x and y have other similar relations); many-one correlations (a father has many sons but a son has one father); and one-one correlations (the order is greater by one).

Unfortunately, functions often are both manifest and latent. Manifest functions are those identified or recognized by participants. Latent functions are those which are neither intended nor recognized.

Effect, the condition of being dependent upon a cause, is always the result of the intersection or combination of two or more previously independent conditions including a functional relation.

OPERATIONALISM

Man does not seek to know or explain just for the sake of knowing and explaining. Scientists and system analysts assume that they desire to use their knowledge for societal or personal advantage. The teleological or purposing nature of man is presupposed.

The purpose of logical explanation is to gain for man the possibility of causal control or operation wherein man can include a functional application in a symbolic or material system in order to achieve a predicted or predetermined state which is an inductive transformation of a prior state. Operationalism provides a way for applying the meaning or utility of predictive logical statements.

Operational capability is fully and systematically present only in cybernetic systems, systems capable of self-initiated and self-controlled purpose and process. When one initiates and performs a specific operation he is moving from the symbolic or verbal domain to the nonverbal or word-referent domain. An operational definition is the instruction for performing a particular task, the observation of which will explain the concept. When a man performs an operation he is inducing change or transformation into a material existence which is the base of his symbolic simulations. He is transforming an event and a system. He is generating new and unique possibilities and probabilities.

As man has advanced in his use of and understanding of symbols, he has discovered the great utility of linguistic speculation. Today theories and hypotheses (new possibilities) are proposed long before their applications can be made useful or operational. Such is the utility of analogy and linguistic speculation, monitored by semi-controlled inference and logical implication. If we require in our affairs that all statements become observable operations before we trust them or regard them as true, the value of such speculation is lost. Thus imagination and intuition will continue to be essential to any significant progress in human-environmental systems. Hypothesis-making or theory-generation tend to precede creation of iconic models in analysis or construction of complex material systems, but not necessarily in simpler or in lesser known (more original) situations. Inductions via creative trial and error experience will maintain their unique value in the latter circumstances.

I want to emphasize once again a singular condition that educators should consider in their efforts to make education highly systematic or scientific. They should understand that any deliberate operation governed by mechanical laws (of logical and scientific method) is, by definition, under partial human control. It is therefore reversible or alterable prior to enactment. Knowledge about (a priori meaning) and know-how (evidential meaning) which add up to *can do* are not sufficient to assure knowledge of *does do* and a posteriori valuing or *should do*. The latter conditions are increasingly within the control of the imaginative, intelligent and informed individual as he initiates and directs his own actions.

The systematic inquirer or system analyst is particularly interested in the material implications of the logical rules of causality. The logical ordering of causal inferences requires the acceptance and maintenance of a number of conditions or rules. If a causes b is inferred or implied $(a \rightarrow b)$, then:

1. a occurs prior in sequence to b.
2. The occurrence of a is always followed by the occurrence of b (there is connexity).
3. If all other factors (c) are held constant, a and b still correlate.

Whenever a correlation between a and b is inferred, three possible explanations exist and must be considered:

1. a causes b.
2. b causes a.
3. c causes a and b.

In the establishing or inferring of causality we should note that:

1. Causes may be necessary.
2. Causes may be sufficient.
3. Causes may be contributory.

Causes that meet the criterion of necessity must always function in order to produce the inferred effect. That which is necessary is true categorically. If sufficient causes are eliminated the effect will be eliminated unless other causes are also operating. Sufficient causes establish partial truths. If a contributory cause is introduced into a known situation it will produce a known (similar) effect.

The thoughtful person is extremely careful in inferring relations of cause and effect. However, Bertrand Russell writes as follows concerning the possibility of proving correlation and/or causality as compared with the difficulty of proving existence:

> As a matter of fact, if any one were anxious to deny altogether that there are such things as universals we should find that we cannot strictly prove that there are such entities as qualities, i.e., the universals represented by adjectives and substantives, whereas we can prove that there must be relations, i.e., the sort of universals generally represented by verbs and prepositions.
>
> (Russell, *Problems of Philosophy*, p. 95)

Russell points out that adjectives, adverbs, and nouns express qualities of particular things and qualitative relationships. Prepositions and verbs express relations between and among two or more things. Mathematics is such a useful language for logical and scientific inquiry because it provides for an almost unlimited number of elements (discrete content or substance words—nouns—including their independently identified quantitative or metrical functions). It also establishes a very limited number of formal or relational symbols to be universal among prepositions and verbs.

THE INTERVAL SCALE

Up to this point we have discussed the nominal and ordinal scales of meaning and judgment and the ideas of logical relation and function. The scale of mathematical importance derivable from the logic of relations and functions is the interval scale which, in turn, is followed by the most refined of all scales—the ratio scale. Summary statements indicating the significant properties of each of the four types of logical-mathematical scales of meaning, judgment, or measurement follow:

> The nominal scale—numbers, letters, words are used as terms for classifying objects or events.
> The ordinal scale—involves determination of similarity and difference, equality and inequality, and rank ordering.
> The interval scale—a scale for determining the equivalence of differences (how much more than, less than).

The ratio scale—a scale for determining exact or absolute proportions of determinate qualities, for ascertaining the equivalences of ratios (how many times).

The scale of greatest meaning for system analysts and educators concerned with relatively ill-defined problems and heuristic problem solving is the interval scale. Interval scales involve these characteristics:

1. The property being designated or evaluated must be defined in unambiguous terms.
2. The property must be such that it can be described in terms of amount (more than, equal to, less than).
3. The property must be such that it can be sampled.
4. The property must be stable enough in the space-time continuum to afford some reliability of measure.
5. The property must be such that an instrument can be devised which will attach numbers to the output effect of an operation automatically.
6. The property must be such that units of it can be established.
7. The measurement of the degree of accuracy or definition of the property must be large compared to the range of variation (error).
8. The interval scale is invariant with respect to order-preserving linear transformations.
9. This scale will provide to properties with the above characteristics an instrument for useful correlation. It will improve the probability of prediction, control, and explanation.

It is important to recognize that the application of number is essential to the developmental use of interval scales. Number, the metrical dimension, is a dependent variable of any function or property. It is an inherent or intrinsic characteristic of that property. This is why we cannot tell what the number of a property is by merely inspecting the property. Functional definition includes number times relation, the intersection of quantity and quality.

THE RATIO SCALE

When *perfect* explanation and control of a system is approached, we are enumerating at the level of ratiocination—categorically true and proportionate reasoning of conclusions from the premises. In applying this level of precise symbolization to the material universe we can say that the problem is extremely well-defined and the solution is algorithmic. We have established a determinate symbolization of part-whole, quality-quantity, and categorically ordered relationships. We can tell and/or foretell how much and how many times in a particular situation a relation is valid. We are functioning at the level of ratio scaling. According to Rapoport (p. 27), the ratio scale is invariant with respect to the similarity transformation.

Ratio scales have these properties:

1. An absolute zero is established.
2. Addition and subtraction of all properties are possible.
3. Any metrical units employed are functionally or practically equal.
4. All units are unambiguously defined and have some logical connection with observable referents or experiences.

In assuming the universal applicability of logical and statistical method that is reflected in the above logical-mathematical system, we assume the mental frame of reference wherein each value is viewed as a variable and each individual variable is a function of another variable, and that it therefore is, or can become, sufficiently describable, predictable, controllable, and explainable.

6

Achieving Precise Representation with Symbols

To progress in inquiry beyond the ascertaining of immediate environmental identities and simple relationships (observed facts), men must employ symbols and instruments or machines to store and manipulate these symbols. He must do this systematically and efficiently. Symbols necessary for orderly thought and communication have to represent our complex and dynamic universe simply yet comprehensively, flexibly yet precisely.

The symbolic systems or languages of logic and mathematics are symbolic structures which facilitate the most comprehensive and precise thinking and communicating. They are terse, yet they have universal application. They have an unlimited number of nouns, but only the few verbs and prepositions necessary for representing the basic logical operations. They transform and/or eliminate the adjectives and adverbs by separating numerical and qualitative functions into number and object event nouns. These languages are ideal for representing via symbols and scales any kind of system and its transformation in the most determinate form or structure presently known to man.

The function of logical and mathematical symbols is to reduce greatly the error of measurement or difference in representation between sign or word and word referents. Careful use of symbols translates controlled inductive experimentation via deductive analysis into near-perfect explanation using interval and ratio numerical scaling. Such progression greatly reduces or substantially overcomes the fallacy of accident either in the sampling of reality or in the logical analysis of its representations. This type of error reduction insures greater human understanding and control.

From Immanuel Kant to Francis Bacon to the present, all scientists advocating the use of the experimental method in systematic inquiry have recognized and advocated the practice of carefully recording each and every element of observation and measurement in their work. Francis Bacon, commenting upon the necessity of instruments for assistance in human understanding wrote: "Neither the naked hand nor the understanding if left to itself can effect much. It is by instruments and helps that the work is done, which are so much wanted for understanding as for the hand."

Niels Bohr, scientific philosopher, explains (in Hempel) the function of logic and mathematics in precise thought and judgment as follows:

> Mathematics is therefore not to be regarded as a special branch of knowledge but rather as a refinement of general language, supplementing it with appropriate tools to represent relations for which ordinary verbal expression is imprecise or too cumbersome.

Words have been found to be quite effective and essential in non-scientific explanations involving nominal and ordinal scaling, but logical and mathematical languages are necessary for the development, application and refinement of meaning attained by representation through interval and ratio scales.

Although any detailed development of Boolean algebra or set theory is beyond the scope of this book, it may be that the reader will be interested in considering a few of the set concepts and symbols and an introductory explanation of their essential character and function.

The fundamental idea of any logical calculus or system is to develop by induction the minimum essential laws; the few assumptions necessary to generate the calculus of classes or sets; and the relations, functions, or operations among elements or sets. From these basic laws the entire system of denotative language and mathematics can be derived by logical processes of implication and inference.

The basic logical relations or structural rules of knowledge of the system include rules for determining equality or inequality—coincidence or lack of coincidence in space-time-function or their representation. Extensions of these relationships lead to determining relationships which are positive or negative and those which are independent or dependent. The rules for conjunction or intersection ($\times$); aggregation, exhaustion of the universe set, or addition ($+$); and exception, mutual exclusion, or subtraction ($-$) can be derived in a similar manner. From these basic logical rules we can develop relational concepts of order among sets (serial, linear, cyclic, spiral, etc.), before and after, betweenness, more or less than, part-whole, and ratio or number relationships.

The algebra of sets, generally referred to as Boolean algebra after the English mathematician, George Boole (1815–1864), is very useful in the generation and communication of logical and mathematical thought. As a logical instrument it can operate with a very simple number system of zero (0) and one (1) when it conceives a system wherein there are only two subsets of the universe: the universe itself (*U*) and the negative or null set (0). An important use of Boolean algebra is in formulating logical categories mathematically for *pure* logic as well as for *applied* logic useful in engineering and in operating digital computers.

The logical rules of systems—that is, the rules for sets and for the transformation, reformation, or reconstruction of sets—can be derived from basic logical relations identified earlier. For example, the basic relations of equality may be expressed as:

1. The reflexive relation: $a = a$.
2. The symmetric relation: If $a = b$, then $b = a$.
3. The transitive relation: If $a = b$ and $b = c$, then $a = c$.
4. The substitution (tautological) axiom: If $a = b$, a may be replaced by b (or b by a).
5–6. The addition and subtraction theorems may be derived from relations three and four above.

The mathematical field laws of closure, commutation, association, identity, inverse, and distribution are derivitives of the above relations, axioms, and theorems.

Venn diagrams and Euler circles are useful for illustrating the way sets may be related. For example, the relation between a universe and a set within the universe is shown in the following Venn diagram depicting an existing set and an empty or null universe (here the set exhausts the universe).

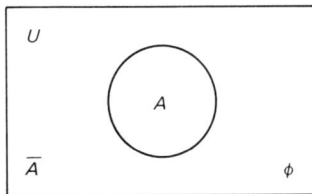

Euler circles are commonly used by logicians and mathematicians for exhibiting the four primary ways in which two sets may be related. These four ways are:

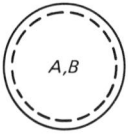

| The identity relation: $A = B$ | The proper subset relation: $A > B$ | The intersection relation: A and B intersect | The disjoint relation: A and B are disjoint |

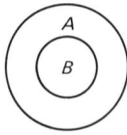

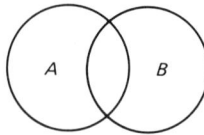

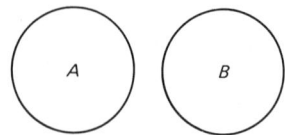

There are three operational concepts fundamental to logical and mathematical reasoning which are applicable to the above four set relations.

The first is the concept of the operation of *election, conjunction,* or *intersection.* The conjunction concept or multiplication involves the elective symbol ($\times$). In an elective operation the order of operation does not affect the result. Nor does repetition of elective operations alter the result either in logic or mathematics ($1 \times 1 \times 1 = 1$). The intersection of sets is the set of all elements which belong to both A and B (inclusive and). This connected intersection is shown in the shaded sections of the Euler circles which follow:

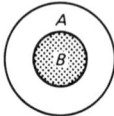

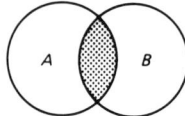

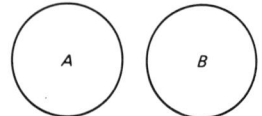

In conjunction, logical implication is true when all components are true and false in every other case. A conjunctive relation is dependent, conditional, and connected. It requires that both antecedents be true. Together they are necessary and sufficient. The negative is also conditional and connected. The logical expression *if and only if* or the double arrow $\longleftrightarrow$ shows this relation ($A \longleftrightarrow B$).

The second concept is the operation of *aggregation, union, independence, nonexclusive disjunction,* or the *nondisjunctive or* is expressed in addition. $A + B$ means either A or B, or, in combination, both. The union of sets A and B is the set of all elements which belong either to A or to B or to both A and B (the either-or relation). The shaded section of the Euler circles which follow show each of these four possible relations. ($A \rightarrow B$)

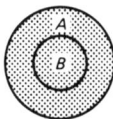

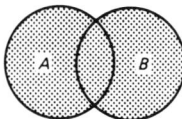

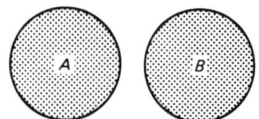

The fourteen laws in the algebra of sets can be derived from the above relations. Nonexclusive disjunction (either-or or possibly both) cannot be true if all components are false. If p, then q ($p \rightarrow q$) or either p or q is generally not completely determined by the truth values of its components. Connection between components of a nonexclusive disjunction is such that it holds if all components could not be false. When one denies p or q one need not deny both components. He only needs to deny that the alternatives have been exhausted. (Fisk, p. 52) For the exclusive disjunctive which is shown below there is an added necessary condition.

The third concept is the operation of *exception* or *subtraction*, represented by the subtraction symbol ($-$). Exception represents a relation of difference, exclusive disjunction, independence (the exclusive or). If two statements or propositions are disjunctive, one must be false and one must be true. In exclusive disjunction, connection holds only if all members could not be false and all members could not be true ($A \longleftrightarrow -B$).

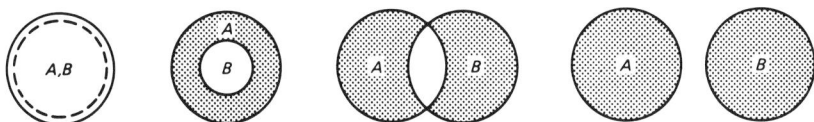

In Boolean algebra the truth of a proposition is indicated by a one (1). Truth and existence are regarded as universals.

The falseness of a proposition is indicated by a zero (0). Nonexistence and/or negation are considered null sets.

In the conjunctive or dependent relation, indicated by $\times$, two propositions are true only if both are true (inclusive and).

The nonexclusive disjunctive or independent relation of two propositions is true if one and/or the other is true. This is indicated by $+$ in Boolean algebra.

The denial of a conjunctive proposition yields a disjunctive proposition (the exclusive or). Since a conjunctive asserts that both conjuncts are true, its denial asserts that at least one of the disjuncts is false. Both of two disjuncts cannot be true. Disjunction asserts the *exclusive or* (mutual exclusion).

The following truth tables represent the possible relations between two propositions when they are either conjunctive or disjunctive:

Truth table for nonexclusive disjunction					*Truth table for conjunction*			
p	q	$p+q$	*Boolean Notation*		p	q	$p \times q$	*Boolean Notation*
1	1	1	$1+1=1$		1	1	1	$1 \times 1 = 1$
1	0	1	$1+0=1$		1	0	0	$1 \times 0 = 0$
0	1	1	$0+1=1$		0	1	0	$0 \times 1 = 0$
0	0	0	$0+0=0$		0	0	0	$0 \times 0 = 0$

Note that in the truth table for conjunction or intersection which represents closed systematic operation (system × event), only when both antecedent elements are true (they exist and the relation is consistent) is the effect or implication true.

Among the many symbols used in Boolean algebra or set theory are the following:

$=$	equal to	$\neq$	not equal to	$\approx$	approximately equal to
$>$	greater than	$<$	less than or	$\geq$	greater than or equal to
f	function		contained in	$\leq$	less than or equal to
$\emptyset$	null set	R	relation	$\longleftrightarrow$ or iff	if and only if
$\rightarrow$	implies	U, V	universe	$\overline{a}$	$-a$ or complement of a
		$\equiv$	is equivalent to		

Logicians follow certain conventions in their use of symbols. Any universal proposition is expressed ordinarily by an equation. A particular proposition, being the contradictory to some universal proposition, is most simply expressed in an inequation. For example (Lewis and Langford, p. 61).

All a is $b = a - b = 0$
Some a is not $b = a - b \neq 0$

A particular proposition asserts an existence. Some a is b means that a's which are b's exist.

$a = b \equiv$ (is equivalent to) $a - b + ab = 0$
$a \neq b \equiv a - b + ab \neq 0$
$a + b = 0 \equiv a = 0$ and $b = 0$

Simple symbolic statements which illustrate the tautological transformation or reclassification from one stated relationship to another include:

All a is $b \equiv ab = a; a - b = 0$.
No a is $b \equiv a - b = a; ab = 0$.
Some a is $b \equiv a - b \neq a; ab \neq 0$.
Some a is not $b \equiv ab \neq a; a - b \neq 0$.

Universal propositions assert a non-existence (of an alternate class).

If $a < b$, $ab = 0$, where $<$ means less than or contained in.
If $a < \overline{b}$, $ab = 0$.
Some a's are $\overline{b}$ (not b's) contradicts $a < b$ or $ab \neq 0$.
$1 = -0$ or $\overline{0}$.
$a \times a = a$.

$a \times b = b \times a.$
$a \times 0 = 0.$
If $a = b$, then $ac = bc$.
If $a = b$, then $a + c = b + c$.
$a = b \equiv a < b$ and $b < a$.
$a - a = 0.$
If $a < 0$, then $a = 0$.
$a - b = 0 \equiv ab = a; a < b$.
$a = b \equiv -a = -b$.
$0 \equiv -1$ or 1
$a + a = a$
$-a + a = 1$
$a + 1 = 1$
$-a + b = 1 \equiv a < b$
$a + b = b \equiv a < b$
$a < 1$ (the universe)
$a (b + c) = ab + ac$
$a + ab = a$
$1 = a + -a$

The negative is represented by $1 - a$, a, or $-a$: $a < b \equiv a (1 - b) = a$
The expression $1 - a$ or $-a$ represents the number of times where the proposition is false.

$a \times b$ or $ab \equiv a$ and b are both true.
$a + b$ means that one of the two is true.
In propositions if $a \neq 0$, then $a = 1$.

Mathematical and logical calculations, however complicated they may be, involve the programing of a series of discrete and simple events or operations similar to those shown. The minimal unit or operation must include two set symbols, a relation sign, and the resultant or dependent equation or total. Ordinary mathematical calculation processes normally involve short-cut algorithms and/or estimates that are much more efficient than the basic Boolean system—providing a digital computer is not available. The most complex reasoning processes involved in Boolean algebra which are capable of being performed by digital computers are simply combinations, intricate as they may be, of the logical categories of true, false, conjunction, and disjunction. Out of these logical operations computers can be engineered to carry out the following operations:

decisions—comparisons of similarity and differences, and subsequent branching of operations;

selection according to some predefined criterion;
checking against a tolerance;
arranging in sequence, which includes matching, merging, tabulating, sorting, implication, denial, shift, and translation.

In addition to their interest in the truth and falsity of propositions (of universals), logicians are also concerned with probability. They are interested in systems that are neither universal nor non-existent. In such systems and their symbolic representations, every implication of truth asserts that a statement is sometimes (sufficiently) true or possible when it is not tautological (necessarily true). (Lewis and Langford, p. 240) Logical probability is a class of arguments in which the conclusions are true with a relatively high frequency when the premises are true.

Logicians employ the concept of implication and the implication symbol of the arrow to illustrate probability or consistency. If $p \rightarrow q$, this means that q is deducible from p. No two propositions (p and q) can be consistent and independent. If p is self-consistent it means that p is possible.

In equivalent or symmetrical statements, $p \rightarrow q$ and its converse, $q \rightarrow p$, are true. The inverse of $p \rightarrow q = \bar{p} \rightarrow \bar{q}$.

In implication, if p, then q, implies $p \rightarrow q$. This is the chief basis for deductive reasoning. This implication is proved false only when p is true and q is false.

If $p \rightarrow q$ and $p \rightarrow \bar{q}$, then p is impossible (is not self-consistent or meaningful.

If $p \leftarrow q$ and $\bar{q}$, then it is necessarily true.

Other symbolic notations for relations of implication include:

1. $p \rightarrow q \longleftrightarrow p$ only if q.
2. $p \rightarrow q \longleftrightarrow p$ is a sufficient condition for q.
3. $p \longleftrightarrow q \longleftrightarrow p$ if and only if q (p iff q). p is a necessary and sufficient condition.
4. $q \rightarrow p \longleftrightarrow p$, if q.
5. $q \rightarrow p \longleftrightarrow p$ is a necessary condition for q.
6. If $p \rightarrow q$ and $q \rightarrow r$, then $p \rightarrow r$. This illustrates transitivity.

The hypothetical syllogism of interest to educators is:

$$p \rightarrow q, q \rightarrow r \quad \therefore \quad p \rightarrow r$$

or

$$p \rightarrow q \quad \therefore \quad (q \rightarrow r) \rightarrow (p \rightarrow r)$$

Shown in the quantified form, it appears as follows (Fisk, p. 81):

$$(x)\,(fx \to gx),\, (x)\,(gx \to hx) \quad \therefore \quad (x)\,(fx \to hx)$$

Of necessity any further explanation of symbolic logic, Boolean algebra, or set theory must be pursued by interested readers in the many other resource texts which are available. The content presented here was included to demonstrate the fundamental extensional function and universal applicability of logical and mathematical symbols. Hopefully, most readers are convinced that logical and statistical reasoning is the essence of advanced systematic inquiry and that system methodology and technology are either direct derivations of or are capable of deduction from interdisciplinary science or system philosophy and theory and general semantics.

In summarizing this introduction to general semantics and psycholinguistics and the role they play in human inquiry and systematic action, it seems appropriate to restate that both semantic content, derived by inductive experience and/or designation, and logical process, involving deduction and analysis, are essential in human efforts to understand and direct events and systems.

Ordinary communications and verbal language are open and flexible. They are the heart of the primitive or subjective scales (the nominal and ordinal scales) of meaning and judgment. They are the instruments necessary for inducing the first consistencies from significant truths.

Logical and mathematical systems represent those symbolic elements which are most effective in simulating or representing relatively closed, complete, and precise operations. They are necessary instruments for advancing inquiry and judgment or measurement toward higher levels of precision and efficiency in systematic classification, prediction, control, and explanation. Please note that nominal and ordinal meanings are assumed prerequisites wherever men apply interval or ratio measurement procedures, however.

General semanticists, psycholinguists, and system experts know and appreciate the need for both flexible symbolic systems and precise logical-mathematical symbolic systems. They understand both the primitively formative or transformational and the systematic or informational functions of these interrelated language uses.

Thus the analysis of language and its functions reinforces other disciplines which assert that man has not and cannot fully explain, control, or reconstruct the significant by the application or the representation of the particular or precise. Deductive analysis and rational behavior are always dependent upon the existence of prior uncontrolled human theoretical and value assumptions. As long as nature transforms itself through accidental or random evolution and man transforms himself and all of the systems he has invented or discovered through the exercise of his peculiarly human capacities for imagination, intuition, speculative and reflective thinking, and invention, there will be uncontrolled variables and restrictions limiting the completeness and precision of systematic inquiry and behavior.

The ill-defined and imprecise problems which consistently or significantly confront men are and will continue to be the problems of greatest concern to humanity. As problems fall under human control and understanding they will be exchanged for or transformed into new problems of greater challenge and significance.

III

SYSTEM PLANNING, ANALYSIS, DESIGN, AND DEVELOPMENT PROCESSES APPLICABLE TO EDUCATION

7

System Analysis

INTRODUCTION

During the last twenty-five years the application of scientific method, coupled with systematic planning, analysis, design, and development, has effected a second technological revolution in the civilized world, a cybercultural revolution. It originated in the early 1940s when groups of our greatest living scientists and business engineers consolidated their efforts and resources in interdisciplinary strategies for developing technologies to expedite the resolution of World War II. Significant advances in planning and engineering procedures evolved with the designing and development during this period of modern battleships, sophisticated aircraft, radar, the atomic bomb, and the computer.

The success of these developments has added a momentum to system analysis and development procedures which seems to increase exponentially. Government interests now employ system analysis in rocket development and guidance, space exploration, national security, war strategy, economic simulation, urban planning, and in almost every other area of critical and costly planning. Business and financial organizations have applied system procedures and technologies in budgeting, economic simulation, automation, routine data processes such as record keeping and accounting, and in most other areas of large-scale planning or evaluation and routine and mass production operations. To the casual observer it appears that system methods and technologies can do almost anything.

As might be expected, the non-engineering members of our society, perhaps a majority of its members, have been confused by the development of these new methodologies and technologies. They would like to understand the basic properties of system procedures, their strengths and weaknesses, benefits and costs. The general capabilities of system processes and their most productive applications are of interest to them. They want to know what system procedures and technologies can and cannot do, when and where to employ them, how and how much to use them. The longer-range social implication of system technologies and theories is their concern—how these instruments can best be employed to assure the more human use of human beings, conserve the natural environment, and provide a better style of human living.

Uncertainty over the emerging capabilities and applications of system processes, confusing even to the experts, is a matter of grave social concern. This uncertainty has been compounded by the entrepreneurial oversell of some theoreticians, manufacturers of both hardware and software, and business technicians, especially since the system technology market has been additionally stimulated by an abundance of private and public research and development money. Thus many non-expert citizens have either been oversold on the immediate utility of system methods and technological products or have overreacted against these new resources and tools.

Educational administrators and teachers, accustomed to operating in an organization with marginal financial resources, have been cautious about committing their resources to untested applications of system theories and methods borrowed uncritically from business management, to the use of obsolescing hardware, and to an occasional dependence upon deficient or non-existent software. This general caution on the part of professional educators during an era of vast technological change has been the subject of considerable criticism.

It is a focal purpose of this book to assist educational administrators and other interested readers in understanding better the capabilities and limits of system theories and technologies and in sensing the direction of future system applications. The emphasis given to explaining system philosophy and theories in addition to discussing procedures and technologies is intended to provide such a frame of reference for the educational leader. The presentation hopefully will be both insightful and practical, serving to identify the fundamental assumptions of science and system engineering, which in turn eventually determine or explain the functional and utilitarian limits of their applications.

It is quite apparent that much confusion concerning system processes currently exists in the United States. Many of the giant system corporations with educational interests, in spite of abundant economic resources, highly expert system theorists and technologists, interdisciplinary staffs of behavioral scientists, and even experienced educational consultants, are retrenching in research and development of some types of technological educational systems. They are experiencing real difficulties in creating systems and conditions feasible for marketing their technologies. At the same time national leaders in politics and education, frequently politically responsive to the business entrepreneur and also frequently ignorant of system procedures, are calling for the rapid development of failsafe technologies and methodologies for mass-educating people efficiently at the very time that inductive learning processes and individualized instructional methods are dominating advanced instruction and progressive educational thought. Some naive system entrepreneurs, accustomed to the simplistic and deterministic thinking that is so productive in the natural sciences and in systematic business operations and marketing procedures, are suggesting that educational system technologies are failsafe, that instructional technologies will replace traditional educators if they do not assume more systematic methods and greater accountability for the learning of their students, and that such technology is equal to live teachers in almost all respects. In some instances the judgments of entrepreneurial pseudo-scientific researchers and technicians are given unquestioned priority over more cautious opinions expressed by educational researchers and teachers in the omni-present debate that seems to accompany educational planning and evaluation at the policy-making levels today. Separation of political or profit-making intent from scientific or systematic content is difficult in such circumstances, where both fundamental human values and significant political and economic values are affected by the decisions.

At the time of this writing it appears that the determination of basic educational policy planning, for a multiplicity of reasons, is to be given more careful and rational consideration. Money for unproductive research and development is less abundant. The cost of innovation and change is dulling the enthusiasm for educational change for its own sake. And the public is serving notice that it will not completely abandon its educational decision-making to either professional educators, politicians, or technologists.

One should caution against overinterpreting the meaning of the present period of conservation and integration, however. The magic power of human imagination, incorporated into an evolving and self-correcting interdisciplinary science and its derived systems of communications and information and stimulated by a general economy of abundance, will never permit society to return completely to the oversimple ultrastability of the past. Thus, the educational leader must either accept his responsibility for keeping informed about constantly evolving developments in knowledge and technology or abandon his role to those who do.

Just what are the particular capabilities of system procedures and technologies? The theories and laws of science and system, including those of communications theory, semantics, and logic, postulate the fundamental but relative uncertainty of events, especially events under at least partial control by quasi-rational human beings. However, they also postulate the almost universal applicability and utility of the methods of science and system, sensing that these utilities will increase rather then diminish in importance as societies, social problems, and the accumulation of human knowledge all become more complex.

System procedures are basically interdisciplinary planning procedures. Anyone who employs recorded information acquired from past experience or learning in order to plan or control present or subsequent actions is employing a system procedure. In such instances goals are tentatively set, experimental or trial events are selectively induced, and error-reducing iterations or retrial events are undertaken. This self-correcting system, eventually integrating multiple goal criteria and complex goal-directed operations in a system, frequently accomplishes all actions and/or establishes all controls necessary for the attainment of the system's predetermined objectives. System procedures therefore are exercised and have a practical value wherever knowledge or information is deliberately applied in solving human problems. They have some utility in attacking almost any human problem, no matter how general and complex or how specific and precise the desired solution must be. The educator can pretty well accept this general assumption.

And educators who accept the general applicability of system procedures can easily deduce that there is no necessary relationship between the application of system procedures and the use of complex mathematical calculations or electronic computers and their peripheral hardware and software. The assumption of such a necessary relationship has actually hindered the educational profession and private citizens from benefiting more fully from applications of less complex and more flexible system concepts and procedures that could be of considerable usefulness in their work.

Several system procedures which are in general use in various kinds of educational planning are explained in the remainder of Part III of this book. Selected for inclusion are those procedures which are currently applied and have been proven to have present utility and further potential for educational administrators and teachers. It appears that these procedures help resolve many recurring or critical educational problems which educators, out of ignorance of these relatively simple or routine processes, might solve less effectively and systematically.

In order to provide educators with a general understanding of system theories and derived methodologies, this section of the book will describe standard macroanalytic and macrodesign procedures (general planning procedures) that have emerged from the experiential and theoretical backgrounds of system oriented scientists and engineers. Among the procedures to be included are a general system analysis and design process called MARS applicable in whole or in part to all planning situations and several adaptations of this model to more specific problems and programs. It will discuss the processes of system planning and development, gaming and simulation, system flow charting, PERT and Critical Path planning, programed budgeting, and operations research.

SYSTEM ANALYSIS

Systems have been defined as interrelated but semi-independent networks of objects and events or the symbols for such assemblies. Events in systems are changes or variations between the input states of elements and their output states, mediated by a function or operation. Events or operations in systems are controlled randomly or inexorably by the forces of nature or by natural forces in combination with accidentally or purposefully induced control of input objects or forces by humans.

Men seek to direct and guide natural forces and objects in order to create causal events in systems, improving and accelerating the attainment of humanly desired output goals or states. In order to facilitate his control over nature, man has invented an elaborate methodology and technology for transforming both his knowledge system and natural objects and forces. He has invented elaborate sets of verbal and numerical symbols for communicating and thinking and sets of rules and methods which organize both the symbolic codes and their transformations into systematic technologies and methodologies. In combination, human systems of symbols and rules of method are so comprehensive and adaptable that all men can employ them to effectively explain, predict, and simulate natural, technological, and living systems, other symbolic systems, or combinations of all of these.

Finally man has combined his observation of natural systems, his formal communications codes, his rules of method, and his technologies to create man-made (machine) models of cybernetic systems, technological systems which control other machines via built-in feedback devices. Thus humanity has accelerated its control over nature almost geometrically, generating simultaneously an abundance of new knowledge about nature, about technology, and even about symbolic codes and rules of method. In the process of mechanizing cybernetic control, men have at least partially automated their self-correcting and quite universally effective method of inquiry, the scientific method. Computers are to a very real degree thinking machines. They can relieve men of many tedious and boring thinking chores just as machines have previously relieved him of much of his difficult, unpleasant, and boring physical labor.

As system analysis is fundamentally a general method of planning for the effective and practical resolution of humanly defined problems, it is extremely useful in organizing the resolution of extremely complex or costly problems. A system analyst uses some type of formal symbolic code and rules of method to organize information which can be used to explain or simulate any known system. System analysis incorporates in its processes the method rules of science, logic and mathematics, economics, and where applicable, politics. It is, however, limited, as is any method, to facilitating the progressive efficiency of function, but it cannot account for the efficiency or inefficiency of dysfunction. In other words, neither science nor system (planning) can explain or create first causes, identify original or unique human needs and goals, or determine ultimate values. Yet systems procedures are remarkable in their ability to expedite the attainment of states of reality closely approximating previously identified first causes and ultimate values.

System processes involve imperfect but controlled induction, either in the primitive form of human intuition and accidental trial or in a more systematic form based upon hypotheses focused by considerable prior experience or theoretical knowledge. Yet, because even primitive inductions are subjected almost immediately to a method of explication, testing, and self-correction, they are progressively and efficiently refined and freed of all kinds of error, including the error of human bias. The system process rapidly and systematically reduces error, eliminating, in order, error associated with the impossible, the improbable, and the impractical. It thereby expedites the attainment of the possible, probable, and practical.

In system literature the generic term system analysis is employed to describe a general planning or macroanalytic process which involves looking at a problem or system organismically as well as analytically. A system analyst is expected to begin his process of solving a problem by first identifying all of the primary states to be achieved in its solution—the program output states. Next he enumerates all of the input elements or resources and all of the output products which characterize the total of all necessary and sufficient operations in the system under consideration. Then he considers what alternatives might be employed in each means-method operation, what inputs, outputs, and operations might be added or subtracted, and how all of these alternatives might be employed singly or in combination to effect the attainment of the total system's program, including its general economic efficiency. He looks at both short-range and long-term effects. Finally he organizes and presents information which explains the technological, economic, and political consequences of using specified alternatives so that the authorities responsible can make more effective and efficient decisions among them. Occasionally he will offer, if requested, a model solution (preferably impersonally mathematically calculated or logically justified) that he believes is superior to the present program or to all other programs under consideration which appear to be less feasible or desirable under present or foreseeable conditions.

Although educational administrators and educational researchers make many decisions which guide subsequent actions in approximately the same manner as the system analyst, there are notable differences between their more intuitive, non-systematic, administrative procedures and standardized experimental or quasi-experimental research methods and the interdisciplinary methods of system analysis to be presented here. The exact nature of these differences will become apparent in later explanations.

As system methods typically embody a series of self-correcting iterations or recycles which progressively eliminate error in the organismic whole or in some component part of a system, they tend to generate large quantities of data. Some sort of simplifying process is needed to make this data manageable. Men traditionally organize, build, test, refine, and standardize plans, models and/or methods for this very purpose. The two principal types of models most frequently employed are the iconic or prototype model (frequently a technological model or machine) and the symbolic model in one of its several forms—a pictorial or graphic model, a verbal-descriptive model, a set of rules of operation (rules of thumb or tested theories), or some type of logical or mathematical equation.

Yet in both system analysis and operations research there is a need for some continuous exercise of intuition and creative thinking. Model building in any form requires either free or controlled induction. In essence all scientific and systematic processes are really pseudo-scientific, as much an art in many instances as a science. However, through the recycling procedure, systematic methods soon sharpen judgment and shape operations by amplifying relevance and expertise, systematically reducing irrelevance and bias.

The main attributes of an excellent system analyst are a tolerance for creativity, ambiguity, risk, and change; an adequate and dynamic philosophy or system for living; an understanding of the methods of science and systems; some facility with logic and mathematics; a productive yet realistic economic and political methodology and sense or intuition; and an appreciation for both the capabilities and limits of method.

System analysis procedures have barely entered the domain of the social sciences or their derivative applied fields such as education. Yet these extremely complex domains or systems are the real focus and challenge of the macroanalyst. It is sometimes said that system analysis applies or "does" philosophy. The epistemology of system analysis is the epistemology of the inexact sciences. Statistics may be used in system analysis, although, if the problems dealt with generally have uncertainties greater than statistical variations within the experimental variables, the extensive use of mathematical statistical techniques is not likely to produce useful results. System analysis particularly emphasizes only those mathematical and logical procedures necessary to account for or control gross uncertainties in extremely complex or ill-defined systems, techniques like testing for sensitivity, use of qualitative ranges such as high, average, low, alternate scenario writing, etc., all basically speculative procedures.

System analysis, as distinguished from operations research, is concerned primarily with extremely complex or ill-defined problems and with the generation and evaluation of alternative programs or problem solutions which are speculative, creative, or prescientific. Operations research generally attempts to develop optimum solutions for rather limited or determinate problems or tasks from among well-known and tested alternatives.

Because of their failure to differentiate between the functions and purposes of system analysis and operations research, professional educators have occasionally ignored the utilitarian value of system analysis and design for resolving many of their ill-defined problems productively. They have occasionally confused and misapplied system analysis and operations research techniques. Perhaps educators are understandably and correctly apprehensive about employing methods of operations research which tend to exploit the premature convergence or standardization of methods for resolving problems. Yet they sometimes do not realize that the very essence of system analysis, particularly macroanalysis, is to diverge or enrich, not converge, economize, and restrict.

The generic term operations research frequently is applied to all microanalysis and microdesign problems—problems involved in developing real-world prototype operations and systems. Mathematical modeling, engineering, and programing techniques are extensions of the basic principles of system analysis to more finite and concrete problems and events. These procedures require in-depth technical training and the use of technical experts and essential technologies for their application. Operations researchers and programers set out to develop detailed plans or programs of prototype or discrete microsystems and subsystems in the form of mathematical statements or equations accounting for or controlling every necessary and sufficient element in the model system. Frequently relying on elaborate management information systems or extensive research, operations researchers and programers employ their program simulations and calculations to produce comparative mathematical indices for direct use in modeling or decision-making. Electronic computers and peripheral equipment are typically employed to accomplish these sophisticated procedures with the requisite speed and precision.

Microanalysis and microdesign of real-world prototype models or their simulation are the basic ingredients of operations research system technology which make it a most valuable extension of natural and acquired human capabilities. Yet the chief values of operations research procedures and technologies are often associated with and generally limited either to extremely critical one-of-a-kind human problems and decision chains or to highly standardized, precisely regulated mass-production programs and systems.

The epistemology of operations research, in contrast to that of systems analysis, is derived from the exact sciences. Operations research emphasizes applied mathematics, using wherever possible and practical linear mathematics and programing, inventory theory, queuing theory, search theory, game theory, and Monte Carlo methods. Operations research thus exemplifies an engineering, microanalytic, or programing process and frequently requires the use of computers with their remarkable memory for detail and their capacity to calculate mathematical solutions rapidly. Operations research imposes arbitrary assumptions rather readily upon objects and events in a system in order to become mathematically precise. It is frequently willing to pay the price of a loss of flexibility in order to achieve lower unit costs or a greater volume of production. Often the degree of control believed to automatically accompany operations research procedures or computer simulation is assumed rather than real, however. This is particularly likely when the model is attempting to represent social systems or systems heavily influenced by quasi-rational human behavior.

In the field of business engineering, where sophisticated system analysis and operations research procedures are most widely used, a system analyst is generally a person who has acquired extensive operations research and programing training and experience but who also has additional training in business and economic theory or engineering. The best business system analyst is the person who has shown an understanding or gift (art) for using this technology in association with other components of the business world, with other machines, record and report systems, and human elements and operations, for the purpose of maximizing the economic production goals of the system. Only the system analyst with advanced theoretical training and a superior political-economic value sense in addition to his programing and/or his applied mathematics skills, is likely to achieve a top policy-making or administrative position in a major business organization.

Perhaps it is appropriate here to point out that education, in contrast to business, involves an organization which is uniquely characterized by its multiplicity of goals and its extremely complex patterns of means-ends objectives. Business traditionally is clearly production oriented; it has a limited and finite number of products, most or all of which are usually accounted for or subordinated to a single economic index figure representing the company's rate of economic growth or margin of profit.

Because of the general applicability of system procedures and their widespread use, system theorists and students have rapidly accumulated knowledge of their basic characteristics and consistencies. It is therefore possible to justify the making of a number of generalizations about their nature. Some of the conclusions system analysts have reached about system methods and technologies follow.

SOME CONCLUSIONS ABOUT SYSTEM METHODS

System analysts seek first to identify the universe of output objectives truly representative of the system's fundamental purposes. Next they identify for consideration each significant input, operation, and mediating output in the normal operation of the system. They know that science reveals that usually there is no single right set of assumptions and values to consider and therefore no single right method for their attainment. Good system analysts will describe alternate sets of assumptions and their implications in such a way that responsible decision-makers can make judgments based upon a relatively full range of relevant information. A good analysis will identify, explicate, and test all probabilities or uncertainties as well as certainties. The minimum contribution of analysis is the identification and calculation of the costs and benefits of reasonably good alternative solutions to small problems (subsystem operations). Better analyses will attempt to estimate the cost-benefit of relatively unknown operations. They will then weigh alternate solutions of each subproblem independently and in combination with all other operations in the system.

System procedures systematically simplify and/or synthesize problem solving operations as well as explicating and analyzing them. They are designed to converge ultimately on effective standardizing or normalizing methods and solutions which prove to be useful in the process of managing or controlling the system's effectiveness. In this desire to simplify, analysts frequently apply inappropriately narrow or "hard" scientific procedures, including the usual operations research techniques, in such a manner that the solutions may have immediate effect but a greater ultimate cost than benefit. Overemphasis upon microanalytic detail and quick economy tend to stress quantitative aspects in systems over their qualitative aspects. System technicians are inclined upon occasion to consider as important only those objectives which can be measured or reduced to numbers (particularly monetary values), thus avoiding or ignoring truly difficult and significant problems.

Within the system methodology, particularly in microanalysis, there is a general and deliberate pattern of goal displacement. Broad subjective goals are first made objective by explicit definition. Operational definitions are next employed to define objective output ends in terms of objective input means. It is assumed that the sum of all objective inputs is equivalent to a qualitative assessment of the total system. In systems with such goal displacement, performance is frequently confused with the system's productive effectiveness. Activity may be valued over product. Maximization of subsystem gains in an incremental and accelerating manner is frequently assumed to insure the optimization of the system as an organic whole. Short-term profit may be valued so highly that long-term growth and satisfaction is ignored. There is a general bias in microanalysis in favor of the old over the new because significant change is hard to justify economically, particularly in the short run. Efficiency, economy, and standardization can become a "path of least resistance" if given sustained priority over long-term growth and health in complex systems such as social systems.

Operations researchers and engineers and computer programers are concerned with developing models which are categorically complete, fully closed, and relatively fail-safe. They seek to create surprise-free programs or methods which will account for every contingency. These programs are progressively effective and efficient as models of mechanical systems or systems with limited and standardized application. But when such detailed and linear programing is applied to the modeling and controlling of extremely complex and dynamic social systems, the system frequently fails. Only social bureaucracy, absolute hierarchy of the worst kind, and utter human boredom can result from such misapplications. The entire energy of the system would have to be consumed to control a healthy social system absolutely or make its future surprise-free or fail-safe.

Comprehensive system analysis (macroanalysis) recognizes that all system procedures involve imperfect information and therefore require induction, including the continuous exercise of an evolving human intuition. All elements or objectives in a system are usually either not known or are subject to change because of changing conditions. The macroanalyst therefore does not usually seek single objectives or single optimum means solutions, nor does he seek exhaustive detail regarding every possible contingency. He often asks the speculative, philosophical, and theoretical *why* questions, attempting to reorder or reorganize the system by generating new goals as well as new means-ends relationships. In his process of macroanalysis he works to expedite both significant and systematic decision-making. He organizes information in such a way as to facilitate the distinction between that which is necessary, crucial, and relevant and that which is inconsistent, insignificant, and irrelevant. He also attempts to account for or predict the effect of judiciously controlling, suppressing, or eliminating unproductive information or operations from the system.

The best analysis will enumerate (make explicit) all relevant means and ends alternatives in such a way that a cost-benefit estimate of their values can be made separately and in combination with all other input-output operations. The optimum goal of all analysis is to propose a plan or program which will account for the continuous and incremental improvement of the total system—its epitomization. The macroanalyst therefore does not frequently seek immediate optimum or fail-safe solutions; this is more the province of the operations researcher. The macroanalyst is more concerned with accelerating the self-correction process, since he is primarily concerned with systems and derivative models which exemplify dynamic growth states rather than static states of equilibrium.

Both macroanalytic and microanalytic techniques are designed to provide a set of rules or criteria for identifying the output value assumptions used in systematically ranking alternative methods and results of their achievement, and for automatically or systematically justifying the selection of an optimum from among better input alternatives. Generally system analysts work cooperatively with other policy-making organizational personnel in determining these explicit criteria since human judgments and values are usually involved. Operations researchers often justify their decision processes more impersonally. They are more dependent upon an internal expertise which is predetermined by the rules of procedure and mechanics involved, and by the constraints and restrictions of prior decision guidelines made by persons higher in the social organization.

Macroanalysts usually present to organizational decision-makers their model criteria for ranking operational alternatives in the form of cost-benefit estimates or lists of positive and negative values. If the model is well done the rules for deciding plus other information will expedite or sharpen the judgment of the decision-makers considerably. In analysis alternatives are judged suitable when they satisfy the requirements set by the predefined goals. They are judged feasible when they can be attained, acquired, or controlled (made possible). They are judged acceptable or effective when the costs associated with them are within reason. They are judged efficient when the cost-benefit index of one alternative master plan is judged optimum in comparison with all other suitable, feasible, and acceptable operations.

In the modeling of social systems, the macroanalyst is aware that the free exercise of human intuition and idealization are essential for keeping such systems productively open, dynamic and healthy. He makes room for serendipity and synergism in his planning of operations and controls in such systems. He encourages the pursuit of serendipity, the happy discovery of valuable or agreeable outputs or products not originally sought. He also encourages systematic synergy, a circumstance where the combined value of output elements or products is perceived to be greater than the expected sum of the effects taken independently (where implicit values and satisfactions can be accommodated as well as explicit objectives attained). Serendipity and synergy are essentially unpremeditated and pre-systematic benefits achieved in a dynamic self-correcting system. They are formative not informative, creative not reconstructive, prescientific not scientific. Macroanalysis which models extremely complex dynamic systems encourages more speculation, more historical or normative human evaluation rather than experimental evaluation, more near-random or idealistic search than does operations research. The latter tends to be efficiently focal in its utilitarian value assumptions and range of operational alternatives.

The interested reader is encouraged to study independently data processing, computer programing, operations research, and system analysis and engineering, the technologies and disciplines used most extensively in system procedures. He can in this way acquire greater knowledge of more sophisticated system procedures, and he may find this degree of investigation and penetration useful, if not absolutely necessary, in advanced scientific inquiry, particularly in the natural sciences or in the management sciences. In all likelihood he will sense that most sophisticated logical and mathematical modeling and applicative technologies are readily justifiable wherever the resultant decisions are critical or where they give a distinct competitive advantage to the original discoverer and/or the most economic mass producer.

8

Basic Theories and Procedures of Gaming and Simulation

Among the many system methodologies and techniques employed in advancing human inquiry and its applications in technology, none are more demonstrative of the full range of basic assumptions and functional limits of system procedures than are gaming and simulation. Gaming and simulation theories and methods clearly indicate that quasi-rational, open and speculative assumptions are frequently the foundation upon which highly rational and closed processes of enumerative and probability mathematical models are construed. It is not strange, therefore, that many of the best philosophical, theoretical, and mathematical minds are focused on these areas of system interests and concerns.

Sophisticated system theorists, system analysts, and social researchers and inquirers are primarily concerned with studying and resolving extremely complex, ill-defined and truly significant human problems. They frequently spend a great portion of their time and energy in speculative and creative consideration of alternate futures of human and natural affairs. Often they are not immediately or directly concerned with transforming their cognitions into effective and efficient social realities as is usually the case with the operations researcher, the educational practitioner, and the technologist or engineer. However, the system processes of gaming and simulation are gradually developing into strategies for facilitating the relatively efficient transformation of highly complex speculations of possible or ideal social-ecological futures into systematic and practical plans for the present and immediate future of the real world. Thus gaming and simulation are of considerable interest to educational researchers and theorists; they are in fact being used in educational administration and instruction today.

Throughout human history there have been two clearly contrasting philosophical and psychological sets toward the solution of ill-defined and complex problems. One approach might be called the imaginative, sportive, open, or gambling set. The other approach might be called the cautious, defensive, or rational approach. The latter approach involves a set toward logical, scientific, systematic behavior. The imaginative individual concentrates on maximum gain with little concern for loss. The rationalist compromises gain in order to reduce risk, loss, or error probabilities. Intelligent human beings are a blending of these over-categorized behavioral sets.

The system theorist can easily perceive the sportive and rational in human behavior blended into a game analogue, wherein all life may be regarded as a game, a contest or series of contests governed by a set of rules which provide a pay-off. The intuitive-activist is the person who plays the game trying to win; for him each individual move is unique and independent of all others. The applied scientist, technician, or engineer tries to develop a system to beat the game consistently; he is the ideal organizational subordinate. The systematic inquirer or theoretician tries to develop a system to beat the system; he tries to improve the game. The speculative or imaginative philosopher tries to develop a system to beat the system to beat the system; he seeks to discover and play a better game. Please note that some mix of sportiveness and rationalism is contained in each type of player listed.

Architect-philosopher Buckminster Fuller emphasizes in his writings the essential characteristic of the conservative-reactionary-rationalistic in human behavior. (Fuller, pp. 31–33) He points out the significance of external environmental conditions upon many major historical changes or transformations in human affairs. According to Fuller, many of man's greatest achievements have originated as massive human reactions (extrinsically motivated) to extreme need-creating conditions, conditions of intense competition including the threat of basic survival. Within this frame of reference, human disequilibrium or reaction often generates a momentum which merges into proactive or self-purposing efforts and accomplishments.

Jose Ortega y Gasset, historian-philosopher, is a leading proponent of the sportive, openly inductive, self-motivated gaming concept of human progress. (Ortega y Gasset, pp 13–40) He suggests that history provides ample evidence that many of the greatest of human accomplishments are sportive in nature, self-motivated efforts directed toward the search for pure nonpurposing entertainment, for play and diversion. Serendipity—the happening upon fortunate discoveries when not in search for them—describes this gaming perspective. This formative and sportive behavior is unique, deliberately unrestricted by goals and normative rules. It is boundary breaking, tending to generate original penetrations of previous behavior and value patterns. Such experiences as evidenced in history are often initiated by persons in positions of relatively unrestricted or absolute power or independence, where their greatest problem is to generate problems (challenges for living).

Ortega y Gasset suggests that sportive events which gradually acquire social value evolve through an ordered sequence of technological classes or levels. He defines the three primary levels of technology as the technology of chance, the technology of craftsmanship (art and strategy), and the technology of the technician (scientific technology). In the technology of the technician, events achieve a state of public acceptance or value which permits broad social dissemination and distribution (reconstruction). Sociologist C. Wright Mills suggests that human social behavior generally functions at the technological level of craftsmanship, art or strategy.

The systematic or scientific-conservative approach to ill-defined problems, events and systems contrasts markedly with the imaginative behavior of the artistic-sportive-creative individual. The latter prefers to make choices and focus intensively on problems involving high risks, whereas the former tends to consider more alternatives than is sometimes necessary and sufficient. One oversimplifies and the other overcomplicates.

Although all human inquiry involves both formative or inductive quest and reconstructive or deductive analysis, there are noted differences in the balance of these processes as they are exercised by particular individuals. Some persons stress imaginative and open behavior almost constantly. Others are cautious and closed. Still other individuals systematically vary these psychological sets in relation to the situation. It is the imaginative and the varied human behavior sets that are the most difficult to simulate with a computer.

The essentials of game theory have been developed and refined by the artistic and philosophical, by the sportive and recreation minded, by the pragmatists and strategists of business and the military, and by the scientific-logical-mathematicians. It is necessary to present some of the contributions of each of these groups of people in order to understand the process and its potential.

Formal and informal recreational games parallel civilization in their development. In their essence games were and are diversionary and motivational. Almost all human institutions have participated in the evolution of games. The family, religious institutions, recreation and sports, business, education, and the military have used games for their institutional maintenance.

War games were perhaps the first games to make the transition from their original diversionary function to an educational or strategic function, stressing the principles of offense, concentration of forces and power, mobility, reconnaissance, probability, risk, etc. (Abt, p. 7)

One might say that gaming is the method of living or inquiring at its fullest and/or most natural (best!). Gaming as a system methodology blends or balances the sportive and the purposive. It combines elements of dramatic conflict, curiosity, direct emotive experience, and role playing (all affective or nonrational processes) with the exercise of strategy, analysis, and prediction (all rational processes). It coordinates the systematic sciences and the dramatic arts. The system approach exercises analysis; the drama creates involvement and motivation.

A game is a social drama or scenario manufactured to involve the players' interests, intelligence, action, and reaction within a shifting system or dynamically changing environment. The system analyst's use of the term *game* implies the involvement of one or more human beings as contestants or active components in a model system. (Rapoport, p. 18) The primary purposes of gaming are to entertain, to educate, and/or to evaluate people. Heuristic programing or systematic trial or search are essential in the designing of games and in the simulation of human and social systems.

Games always involve multiple input and control centers, at least two dynamic and semi-independent self-regulating input elements. Their systematic relationships are not elegantly simple and calculable. This is the kind of environment which is an analogue of human and social systems.In such environments the rules of the game are slowly modified by the players. Games are evolving, transformative, and incremental; thus they can be used to model effectively social systems containing semi-independent cybernetic elements (people).

When games are designed primarily for entertainment, education assumes a lesser instrumental value. In educational gaming, entertainment is the instrumental or output value compromised to some degree.

The theory of games involves a number of systems of classification. One such system classifies games as showdown, strategy, or combination games. (Abt, p. 6) In showdown games each player exhibits his best performance and fortune independently (without interference). Races, poker, and golf are showdown games. In strategic games players interfere or compete with one another. Bridge, boxing, and chess are games of strategy. In combination games there are often strategic exchanges preliminary to a showdown. Common combination games include football and hockey.

Whether for recreation or for education, games may emphasize the elements of skill, chance, reality, simulation, or fantasy in their effect upon players. As instruments for educating, skill games emphasize the capabilities of players. They reward achievement and encourage individual risk with responsibility. They often discourage slow learners, dramatize player inequalities, and feed the ego of the persons possessing the player advantage.

Games of chance demonstrate the limits of skill and effort, thereby tending to encourage underachievers. (Abt, p. 14) They are often popular in poverty areas. They tend to humble overachievers, minimize personal responsibility and skill. They also tend to induce passivity or magical (idealistic or unrealistic) thinking.

Games of fantasy such as dancing or skiing are emotionally refreshing and stimulating. They are generally highly motivating and possess low cognitive content. These games possess the positive value of releasing players from traditional perceptions and inhibitions.

General simulation games or games of reality have a long tradition. They simulate the nonplay real world and are widely represented in literary fiction, television, and the theater. *Monopoly* is a formal reality game. General simulation or reality games for children or students are educational in that they demonstrate structural relationships of systems and the problems and motives of others. They permit vicarious experiencing (role playing) which is normally beyond the limit of direct possibility. Children's simulation games motivate as they capitalize on the child's longing for adult reality. Most games for children are strategic games and reality simulations rather than showdown games, although they do not especially minimize skill, chance, or fantasy.

In traditional educational methods problems, demonstrations, projects, and case studies are isolated, partial, or incomplete gaming processes. More advanced gaming and simulation processes are now being developed for use by social studies teachers seeking better ways for teaching political and social sensitivities, strategies, and values. Educational games tend to become games of skill with practice.

There is a general danger of teaching oversimplified analogies in game methodology, and games tend to overrate the determinate character or predictability of events in an extremely complex system.

Good games can accelerate the learning of many levels of cognitive and affective objectives. They can simulate, teach strategy, teach the calculation of costs and benefits, risks and rewards, and the weighing of alternatives. Communicative, persuasive, and negotiative skills are developed through games. Games generally involve multiple-sensory experiences and a combination of emotions and cognitions. The values of loyalty and cooperation and the limits of competitive rivalry, greed, and neglect can be demonstrated in games. Games can be designed to punish in a non-fatal and non-personal way either overaggression or apathy.

Viewed as educational or inquiry methodologies, games present simultaneously progressing multiple interactions (complex events) which are first examined one at a time and then gradually telescoped together to achieve integrated yet comprehensible activity.

The distinct advantage of all games lies in the affective domain. Games, used properly, can increase motivation. This very attractiveness is also a real educational or inquiry handicap. Attractiveness and enthusiasm alone are not substitutes for intellectually skilled behaviors. Emotions fundamentally cause people to overgeneralize or oversimplify.

GENERAL SIMULATIONS DESIGN

Whether for the high school social studies teacher, the professor of educational administration and supervision, or the trainer of leaders in business, politics, or the military, the general simulation game has potential with or without computerization. In fact, wherever interaction training is of primary importance, active participants and contestants are essential.

Simulations design begins with the translation of an analytic model of a social system into a game which can communicate the results of operations and decisions by players and the implications of these results to the players. Psycho-social, socio-political, and socio-economic system decisions and operations make excellent gaming content.

The plot or design of a game is a simplification of an analytic model. It mixes analytic truth and dramatic communication. Game design requires that decisions be made concerning which subplots, characters, and events most lucidly dramatize the material to be conveyed. The game purposes, player objectives, allowable activities, win/lose criteria, team patterns, and other rules are then developed to achieve the intended purpose. (Rapoport, pp. 19–20)

Gaming and simulation design involve many compromises or trade-offs between the competing objectives of comprehensive realism and simplification for the sake of playability. Heuristic programing of games considers the desirability of trade-offs between realism (at a cost of ease of playing) and simplification (at a cost of sufficient learning). Concentration (at a cost of topical coverage) must be traded with comprehensiveness (at a cost of detail, realism, and game impact). Dramatic motivation (at the cost of calm analysis) must be traded with analytic calm (at the cost of reduced involvement). (Rapoport, pp. 12–13) The balance, variety, and order of elements of the type mentioned should be determined by the game purposes and its tested effectiveness.

In order to maintain an acceptable level of stimulation as well as simulation, games must be simple to learn, easy to administer, not too subtle in what they teach, and basically interesting to the players. General criteria for game and simulation design include:

1. Games should be easy to learn but not so easy to master.
2. Games should be used intensively, in a concentrated manner, but not overused.
3. Games should fit the purpose and the players.
4. Games should fit particular situations and facilities.
5. Game presentations should involve some showmanship and organizational skill.
6. The game goals or lessons should be fairly obvious.
7. The players should both enjoy and respect the games.
8. Good simulators may not be good stimulators; simulation requires systematic analysis as well as good dramatic plotting.

The blend of effective and cognitive learning taught through games and simulation has the capability of creating desires for changing ways of thinking and acting. Games simultaneously motivate learning and provide processes for change.

It should be clear to the reader that gaming and simulation design, whether non-computerized or computerized, is not necessarily inexpensive or easy to accomplish. Some computer games may involve years of preparation and refinement (tuning or debugging). In fact, much of the value of game design comes in the learning achieved through this preparation. The systematic analysis required and the plotting of the games involve skill and concentration—deep penetration into the system being simulated. Often such preparation requires extensive communications between informed theorists and experienced practicioners, bridging the gap between these groups and educating them.

The professional educator reading this material on game design must have by now noted the similarity between the description of gaming and simulation and the principles of good curriculum design and teaching. They are essentially equivalent.

SYSTEM ANALYSIS FOR GAMING AND SIMULATION

Advanced mathematical and experimental game theory and simulation begin where the generalized design description just presented ends. They are oriented primarily toward reality simulations and usually involve a complex pattern of cognitions and purposes. Motivation and affective elements are of interest only where they relate to purposes. Mathematical game theory is to games of strategy what probability theory is to games of chance.

Systematic game design begins with a comprehensive logical or mathematical model of a multidynamic system, including the induced purposes of the system. It simulates the subplots or designs of human player actions and decisions. And it requires heuristic testing and refinement of the original design model together with abstracted humanoid simulation through repeated test playing.

In the preparation of the model, specifications of the substantive scope (rules), elements and factual details (rules), and the relationships or operations (more rules) are made. A problem space or plot is selected. This is a situation or series of situations involving multiple interactions. The major decision points and the status of information reaching these points are determined. The major operation or communications points between subsystems or external systems elements (inputs) are selected. Typical external systems involved with organizations are individuals, other organizations, and the public.

The order of sequence and the flow rate of the plot or simulation are established. In general, a cyclical and cumulative pattern of problem sequencing is used.

The determining and spacing of points requiring decisions is fundamental to gaming and simulation. These points should reflect and reveal to the players the structure of the social system and events being simulated. Primary conditions of decision points include their multiple interaction or interface dimensions (their domain and range) and their negotiations or bargaining properties (whether they possess conflict or cooperation alternatives).

In complex and dynamic organizations most decisions or problem resolutions are not simple. A specified decision or action might lead to a number of outcomes, depending upon situational factors. In many cases a decision is determined not by personal choice but by someone else's choice of action. Games are distinguished from non-games or determinate situations primarily by whether the choice of actions and certain outcomes can be unambiguously defined, whether their consequences can be predicted, and whether the choice makers have distinct preferences among alternatives and/or outcomes.

Many of the decision rules of games specify or build in behavioral sets or orientations such as cooperation, aspiration, bias, slack, pressure, non-communication, etc. Empirical rules-of-thumb are often considerations in determining the specification of operating and decision rules in organizational simulations.

Games may be set up as supersilent, silent, or noisy simulations. (Shubik in Cooper et al., pp. 449–463) This terminology relates to the amount of information feedback available to a player as to the effect of his or an opponent's moves. In supersilent games, players learn nothing until the game ends. In silent games, a player gets periodic knowledge of his status as compared with his competition. In noisy games, players receive considerable feedback from their opponents and/or their teammates. They have an updated resume of all prior moves.

A decision analysis preparatory to game design usually identifies the relatively stable, normative, or consistent criteria or motives of decision-making entities in a gaming situation and orders them into patterns or categories of values. Product goals are the high priority values; process goals are of lower priority. This analysis also determines the normal and/or exceptional perceptions of these rules by organizational participants. Problems arise when there are discrepancies between role perceptions of either the ideal or the actual situational conditions. Payoff criteria in gaming usually include a reward for discovering the saliency or obviousness of the system's decision-making or value patterns.

Martin Shubik has stated that there are in existence three broad divisions of experimental gaming.

1. Games may be concerned primarily with learning, problem solving, and organization (one-person nonzero-sum games of interest to psychologists and sociologists). The player competes against chance and/or his own record.
2. Games may involve two or more teams in face-to-face communication where threats may be made and bargaining, haggling, ploying, etc. may take place (of interest to labor negotiators, international politicians, economists, and social-psychologists).
3. Games may involve two or more teams where there is no direct communication (of interest to economists studying the market).

According to Shubik there are many solution concepts in gaming. These include non-cooperative equilibrium, beating the average, efficient point, market share, price leader, aspiration level, dynamic equilibrium, and games of economic survival. There are an equal number of technical solution concepts with similar esoteric titles.

Shubik points out that game theory and simulation are primarily methodologies and cannot be expected to be substitutes for intellectual disciplinary research or instructional methods of other types. He notes that game theory has a limitation in that it imposes severe *noise control* on the simulated environment. It artificially closes an indeterminate system. Systems with noisy environments (adaptively purposing and semi-rational systems) are extremely hard to model mathematically. Many games are too simple and attempt to control strategic choices too closely.

There is a need for games which simulate environment-rich situations such as an educational organization transacting in a public environment.

Some games involve cooperation in order to maximize and divide the gains. Information games and simulations have been designed to test organizational structure and design, i.e. simple versus complex organizations, etc. They can test perfect rationality versus undervaluing or overvaluing one's own organization or that of the competition. They can test the effects of motivation. Games test interaction potential and the effects of limited interaction. Production exercises can be simulated, including buying, maintaining personnel, production operations, storage, distribution, profit, etc. Most business simulations have provided rather limited alternative decision choices. Real innovations have not been too frequently applied or produced via gaming.

Modern computers are excellent for simulating complex system models and environmental dynamics. They are much less expensive to operate than test plants or models are to build. Computer simulations can include fast-time simulations (predicting runs) which permit a determination of efficiency in systems rather quickly. They are useful for modeling natural or mechanical determinate systems in operation, but they do not simulate human behavior well.

Real-time simulations involve human operators participating at real or near real-time rates. Computer real-time simulations involve systems and human task analyses prior to simulation.

George Briggs (in Cooper et al., pp. 479–492) proposes these criteria for justifying a particular type of simulation:

1. There must be a fidelity of simulation. This includes a fidelity of human function, a face validity.
2. There must be a set of semi-independent instruments or measures of a system's performance.
3. Ease and simplicity of control should always be a consideration.
4. In a simulator reliability is inversely related to variability.
5. A good research tool is flexible in the sense that a variety of experimental conditions can be implemented without a major readjustment of the basic simulator.

The perfectly systematized logical or rational instrument, the computer, simulates or exercises its magic when it attacks complex but necessarily very precisely defined problems, events, and systems. It requires the categorical specification of inputs, operations, and outputs. Even the heuristic or search capabilities of systematic simulators must have their domain and range of search or branching operations limited or specified. In spite of their speed and precision, mechanical simulators require a human to tell them when, where, and how to start and stop, to periodically transform or regraduate, or to reorder the valuing system.

The concept of simulation is closed, reconstructive, and informational in nature. It seeks to completely program or predict and then recreate reality. Its ultimate refinement is in the creation of perfect inductions, in which the simulator is guided by exact mathematical model programs which isomorphically simulate the real world of the system.

Computer involvement in gaming and simulation should be realistic. In games designed to educate the players it should be only a segment of the game methodology. For many games designed to socialize, direct player interaction is essential. A key element in executive success is the ability to present and justify ideas to others. Simulations training leaders should exercise human communication and interaction.

When games are designed to train participants, boards of judges can often assist player participants in improving their strategies, clarifying the possible alternatives and the limits of present strategies.

The use of games for testing and evaluation is of questionable value. Testing games lose much of their basic motivational value. And evaluative games cannot be validated until the criteria for model players are known and all cause-effect relations of decisions completely determined.

Experimental simulation methodologies per se serve four major purposes: They train research analysts. They give the experimenter some indication as to how his model would work in practice. The economic simulator is useful for controlling or monitoring a system under study. And finally, simulators can be used to forecast the long-run effects of alternative policies once an experimenter is convinced that his simulator is valid and reliable. Because of these capabilities, economic simulation is one of the types of simulation best adapted to the computer.

The history and practice of experimental gaming and simulation is beginning to reveal the a posteriori values of these methodologies to interested educators. We know that real or typical organizational situations often provide less information and more complex dynamics than simulated situations. Effective simulation games should provide a realistic number of cues, almost always requiring probing by the players for additional information.

The measure of a good game is not the number of decisions required but the number of kinds of decisions a team should make. Competing or interfering decisions, buffering or delaying decisions, and cooperating or facilitating decisions should be intermixed. The game world or universe should be complex enough to be exciting but simple enough to be realistic. It must be shifting and dynamic.

Games for leadership training should facilitate exploration, search, probing, questioning, and analyzing prior to making assumptions or hypotheses of solutions to problems (decisions). They should improve goal-setting ability. They should expand organizing abilities, the discovery of relations. They should precipitate or generate responsibility. And, they should improve the players' abilities to assess and classify their experiences.

In reconstructing models of organizations, simulation has generally concerned itself with the physical flow of goods, materials, and funds. Simulations can be developed around the properties of information, matter, and energy as well. Business firms are often simulated as a set or series of decision centers (positions) each accompanied by a set of decision rules. Each decision rule has a program specifying how the decision is to be made.

Although many significant goal concepts of organizations such as cooperation, competition, and collusion are not well enough defined or stable enough to be useful in computer simulations, certain of these qualities can be built into the simulation. In simulations with live actors, an analyst can observe the range and diversity of such complex and varied operations.

The works of Anatol Rapoport, Martin Shubik and many other authors are available for a more penetrating introduction into mathematical gaming and simulation. The literature on gaming is voluminous and there are literally hundreds of management and leadership games in existence.

Many of the theoretical contributions of systematic simulation lie in the area of systems of classification of elements, properties, and concepts. This is typical of all sciences in their early stages of development.

Games are classified by the number of contestants—two-person, three-person, n-person, etc. They are also classified according to the amount of information available to the players. Games in which all choices of all the players are known to everyone as they are made are called games of perfect information.

Games are classified as to their zero-sum or nonzero-sum status. Zero-sum games are games in a completely closed system. They are games of pure competition wherein one player's gain is another player's loss. Distributed or zero-sum bargaining resolves such competition or conflict only by concession or compromise. In zero-sum games with semi-perfect information distribution and with the advantaged player clearly distinguishable, there is almost zero motivation to continue the game. This game situation approximates the organizational situation wherein the reward, the product goals, and even the methodological goals and decisions are determined unilaterally in an authoritarian or prescriptive manner. In such a situation, cooperative decision-making is absolutely essential to encourage any self-motivated participant to play the game.

Nonzero-sum games are games in which cooperation and collusion can facilitate favorable profit for both parties. In games permitting co-operation there is usually a mixture of commitment, negotiated agreement or cooperation, and competition or nonnegotiable behavior. This is typical of semi-independent, self-motivated social systems where the commitment is always of a limited nature. In real life, agreements are often unilaterally imposed and nonnegotiable, but this power situation is frequently countermanded by contestant reaction preventing enforcement or continuation of the commitments. The teacher strike is an example of such a situation.

In game theory ethical behavior is defined as carrying out intentions considered good nonzero-sum strategy (cooperation) and ignoring the possibility of the opponent's taking advantage of one's good intentions. This definition may have significance in real life!

Game theorists are repeatedly discovering that the zero-sum competitive game is fun as entertainment but is not realistic, productively practical, or cumulatively motivational in real life. The American habit of overvaluing competition is thus called to question by the results of experimental simulations.

The processes of logic are fundamental to mathematical game theory. Specified and controlled inductions (nonzero-sum goals) are converted into process-recursive, deductively analyzed game trees or matrices which exhaust all possible decision alternatives and assign exact payoffs or weights to each branch of the tree. Perfect enumeration naturally permits making the best decision for the situation.

Actually the experience of games of strategy teaches us that the perception of individual or social human behavior as closed or purely rational is a ridiculous and pathetic perception of reality. Conversely, the probability of making ideal decisions in social systems is low. The situational context in extremely complex systems often requires utilitarian, optimizing, or appropriate decisions rather than epitomizing decisions.

It is beyond the scope of this book to pursue the theory and methodology of gaming further. Much literature is available on the subject. The mathematical game theorist, like the system analyst, seeks to develop a logic of strategic systems. The operations researcher and computer simulator often seek to model extremely complex cybernetic systems. The applied scientists and practicing educators are trying to apply the logic of gaming and simulation to the processes of motivating and educating people in a social context.

9

System Planning and Scheduling
Procedures

FLOWCHARTING AND PERT/CRITICAL PATH

Among the useful applications of system methods to the solution of managerial problems in business and government are several standard techniques employed in planning, charting, and scheduling operations in an organization. System analysts and computer programers have modified traditional flowcharting procedures to include all components essential for careful system analysis and design. In addition, system planners have developed standardized symbols and charting procedures for (1) system flowcharting of general operations, data processing systems and equipment, and (2) program flowcharting of planned computer programs. These two general classes of flowcharting techniques are applied extensively and systematically. System flowcharting in particular is a very useful and generally applicable means of presenting a pictorial or graphic model of the major steps in complex networks of simultaneous or sequential operations.

The technique of system flowcharting is described and demonstrated below, employing a simplified charting system which adapts the procedure to general educational use. A little further along a special kind of scheduling procedure employed widely for planning, scheduling, and monitoring one-of-a-kind operations of a critical or costly nature is presented. PERT (Performance Evaluation and Review Technique) models which include the concept of *critical path* are presented and explained as examples of this kind of system procedure.

117

System Flowcharting

A system flowchart is a pictorial graph illustrating the sequential flow of information and/or action events in a system. In system flowcharting, a technologically workable system output goal is specified prior to the scheduling or sequencing procedure. Then the system analyst or administrative planner develops a carefully ordered series of tasks and events for optimally achieving that goal.

An important component of system flowcharting is the determination of necessary decision points for the human monitoring or reviewing of conditions in all operations and subsystem elements up to that point in the process. Terminal decision points plus feedback loops following events permit a recycling or error correcting process which provides for additional trials or corrections before further progress along the operational sequence is allowed. In flowcharting, each system goal is decomposed into a functionally ordered series of tasks controlled by essential decision points and feedback loops.

In the process of flowcharting, the procedure first requires the specification of all necessary input elements, particularly the exact specification of the focal or composite system goal or goals. Their exact conditions or requirements (including numerical incidence) are specified in order to determine both initial and ultimate acceptance or rejection. From that point on, the line of flow indicates the point of entry of all inputs into the system and the quality and quantity of operational procedures and/or informational modes to be utilized in transforming inputs into the desired output state(s).

In human task analysis, the nature of the physical and intellectual activities required by human operators or actors and the characteristics of the output devices employed in the operations (the machines, instruments, techniques, and methodologies involved) must be specified. Human task and event analysis in particular must include a prediction of operational precision or efficiency based upon reference to norms of prior performance measures.

Computer program flowcharts require much greater detail than system flowcharts which involve human acts intersecting and monitoring machine processes. Program flowcharts require a categorically complete statement of all sequenced operations and decision points in the application of the program to the stored data being processed. Computer programing may be explained schematically as involving the following elements:

1. A collection of input elements which are acceptable and whose

exact movements through the system in a series of events are predetermined (known).

2. Ordered submodels or events are predefined. They include all transactions of the input elements with each other or with the remainder of the system.

3. The submodel or event subsystems are connected in specific manners by input-output flow lines (feed forward) and by feedback lines connected to the control or monitoring system.

4. At each designated time pulse (determining exact space-time-function coincidence), the sequence of subsystem operations are induced in the prescribed order. They are subjected to the pre-specified logical rules, the predefined process controls.

5. At the end of the cycle all submodel or event outputs being processed in the system are transferred to the next points of input, the elements are moved ahead according to the systems passage or decision rules, the systems registers are updated, and the cycle is repeated. This process continues until a prearranged stop signal is received. The results of the entire simulation or operation are then displayed and evaluated.

In its finished form a computer program flowchart provides:

a graphic picture of the problem solution
a graphic record of program logic used for coding, desk checking and
   debugging while testing
verification that all conditions possible have been considered
documentation of all aspects of the program.

The degree of precision involved in information inputs and the detail required for presentation of information in computer program flowcharts is beyond the scope and purpose of this book. It is reflected in the highly standardized and exact computer languages employed in programing. Interested readers are referred to the many written sources available on the subject.

System flowcharting, on the other hand, whether machine-oriented or general, is perhaps of greater interest and value to educational administrators. It is possible to describe the elements of system flowcharting technique in such a way that an administrator inexperienced in flowcharting may employ the technique in his planning work. System flow-charting is an intermediate stage of analysis and graphic display, somewhere between normal human operations and precise computer programing or operations engineering, processes involving exact logical-mathematical models. As compared to mathematical models or computer programs, system flowcharts are less prescriptive and less precise but more generally applicable.

The value of system programing and flowcharting, just as with system analysis, lies in charting less well-defined and controlled systems and events. Any flowcharting procedure is valuable (1) as an input planning procedure requiring careful analysis of the relationships of a number of elements and events to each other, and (2) as a readily accessible output document for recording or storing and distributing or communicating the essence of such a system of operations.

A system flowchart may be used to describe the order of physical operations in a system or it may describe the flow of symbolic data and related machine data processing operations in a system. It may indicate material operations or the symbolic control information flow representing or simulating them.

Standard manuals of system flowcharting involve three basic flow-charting symbols: the input/output symbol, the process or operation symbol, and the flow direction symbol (a line with arrows indicating the direction of flow). In addition, a standard IBM template (see IBM Manual listed in bibliography) contains fourteen other system flowchart symbols representative of other data processing equipment operations and elements.

Computer flowcharting employs the three basic system symbols, six special programing symbols, and the fourteen supplementary systems symbols where they are applicable.

A minimum number of the standard flowcharting symbols, eight in all, are employed in the demonstration of systems flowcharting which follows. Symbols indicative of special machine operations are omitted from this presentation as are most of the programing symbols. Readers are referred to manuals and texts by IBM and other authorities for a more detailed course in flowcharting.

The three primary system flowcharting symbols are:

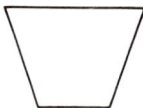

Symbol for *Input/Output*	*Symbol for* *Operation*	*Symbol for* *Flow Direction*
Any type of input/output data or medium	Any discrete event, task, or operation	Flow direction is normally from left to right, top to bottom

Five additional symbols of general value in constructing meaningful system flowcharts include the following:

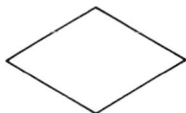

Decision

A point at which a
human choice or branch
is available or required, a
point of alternation or recycle

Predefined Process

The state of a primary
system or operator at the
point of intersect. Prior
detail is omitted

Keying Operation

The state of an operand
(secondary subsystem to
be transformed) at the
point of intersect

Program Modification

A point at which a program output
and/or process goal may be changed

Terminal

An arbitrary point from which
to begin or end a system program

In this book these eight symbols comprise the set of symbols employed to demonstrate the utility of flowcharting for describing systems and events. In addition to the unique form of each symbol, a flowchartist often adds further identification to each symbol on the chart by adding a word title or specific description of the element or operation it represents. As the system flowcharting explanation being presented here is introductory and general in nature, the symbols selected are defined in rather broad or connotative dimensions. They are not intended to represent machine processes as most system flowchart symbols are. For the purposes of general flowcharting the stipulated definitions of the above eight symbols are as follows:

The *input/output* symbol represents focal symbolic elements, goals, problems, or ideas introduced as inputs and produced as outputs. They represent initial and final information elements in the system.

The *operation* symbol represents any event, operation, process, or program in the system. This widely used basic symbol is often employed almost exclusively to indicate an ordered series of discrete operations.

Flow direction symbols are usually indicated by a solid line. As such it indicates feedforward or progressive and incremental action. Operational flow is normally shown as progressing from left to right or top to bottom on a page. Arrows are customarily imposed on flow lines which oppose this normal direction. Broken lines with arrows are normally used to indicate feedback, information flowing back to the cybernetic control system. This information feedback system contains signals generated by monitoring or evaluating operations which return to the operator to guide him in the continuation or modification of his processes.

Decision symbols are used to indicate important points in a system where alternative actions or choices are possible. They usually represent a point at which the control system can constructively monitor the operational flow system. Decision points should be introduced in systems prior to and subsequent to major events or operations. The system should monitor each important event to note the effect upon the secondary subsystem being deliberately transformed and also the condition of the primary or operating system following a major operation.

The *predefined process* symbol normally represents a group of operations not detailed in a particular flowchart. In this presentation it is used to indicate the state of any primary subsystem or operator at the point of intersection with the system being charted.

A *keying operation* symbol represents an operation ordinarily using a key-driven device. As employed here it signifies the state of any secondary subsystem or operand (a subsystem being purposefully changed) at the point of intersect.

The *program modification* symbol indicates a point at which an instruction or group of instructions changes the program itself. The text of the symbol should indicate the nature and purpose of the modification performed.

Terminal symbols represent points at which a program or predefined series of operations originates or terminates. As terminals are arbitrary points which usefully identify or categorize elements, events, or systems, they optimally should represent points of natural or clear definition. Thus they often indicate states of nature or natural equilibrium prior to or following a controlled operation.

In the design and development of a system flowchart, an analyst displays all significant elements and operations. His program establishes and reflects a natural or categorical order of flow among all major operations as they apply to effect a predetermined series of transformations upon an operand or object to be changed. It should also indicate the articulation of all primary system operators and operations in respect to the particular system's total activities and clearly indicate all decision points where control monitoring and/or decision-making is necessary. This level of flowcharting is essentially prerequisite to the development of more detailed computer programing flowcharts or mathematical-logical models of operations.

In any system process, the system can be said to be categorically ordered only when it has prescribed a linear or simultaneous order among all states or conditions and the operations necessary for achieving the prescribed order of states. Any highly rationalized or efficiently planned system must categorically specify its ultimate goals of operation and a serial or simultaneous order of production events which meet the predefined and measured change states expected at an event series terminal point, a point of feedback or decision-making. Each step or state of flow indicates successive and efficient progress from an original condition to an incrementally higher priority output state.

In order to assist the reader in comprehending the essential elements of system flowcharting and the versatility of this technique, several very simple system flowcharts are presented below. The first figure illustrates the basic properties of a single event or operation in a purpose-oriented cybernetic system. A single causal event becomes determinate only when the conditions of its inputs are such that the event transaction is a completely sufficient mediating or penultimate (next to last) condition for achieving the desired output state. In such a situation the operation itself is equivalent to the goal being achieved. It is categorical and algorithmic. Its iteration or exercise always produces the desired effect.

Heuristic events are operations which have a degree of probability considered less than certain in their cause-effect relationships. A predefined heuristic program will solve a particular problem or produce a particular effect only if the sum of all a priori conditions (the series of trials) are sufficiently favorable and eventually account for the penultimate condition.

Another simple system flow diagram shown in Fig. 3 illustrates the basic elements and operations involved in accomplishing a specified transformation in an educational system. This illustration expands upon or details the planning and evaluative possibilities in such a system. The chart in Fig. 4 does not detail a series of operations but lumps all requisite operational events into a single process symbol.

The third illustrative flow chart, (Fig. 5), indicates the importance and frequency of decision monitoring in a highly dynamic or interacting system. This diagram depicts graphically the heuristic problem solution for crossing a street to the corner diagonally across from the initial point of entry. An insert on this chart indicates the gross oversimplification which a casual observer might employ to graphically model this event. The system flow chart depicting the street crossing is in fact very much of a simplification of the actual operating and monitoring program itself. A computer program or logical data program illustrating such an event would be immeasurably more complicated than the system diagram presented.

Still another figure, (Fig. 6), presents a graphic model of a master scheduling operation of interest and concern to many high school principals. The system symbols presented earlier are augmented by the addition of symbols representing documents and punched cards. The opera-

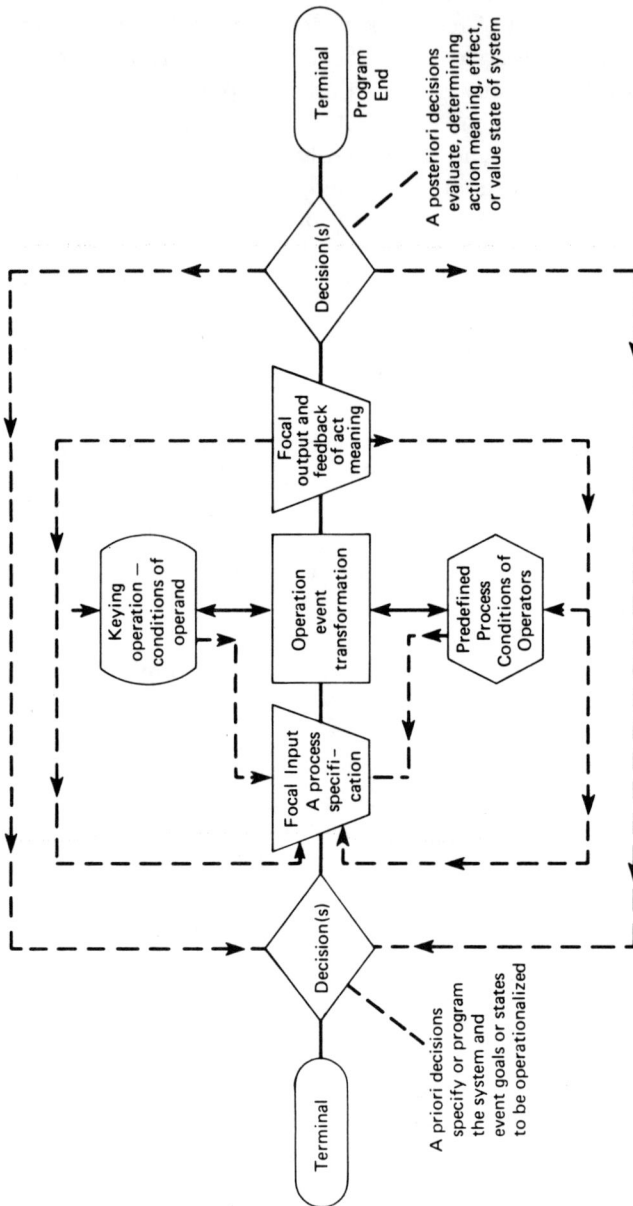

FIG. 3. Flow chart of an event in a system.

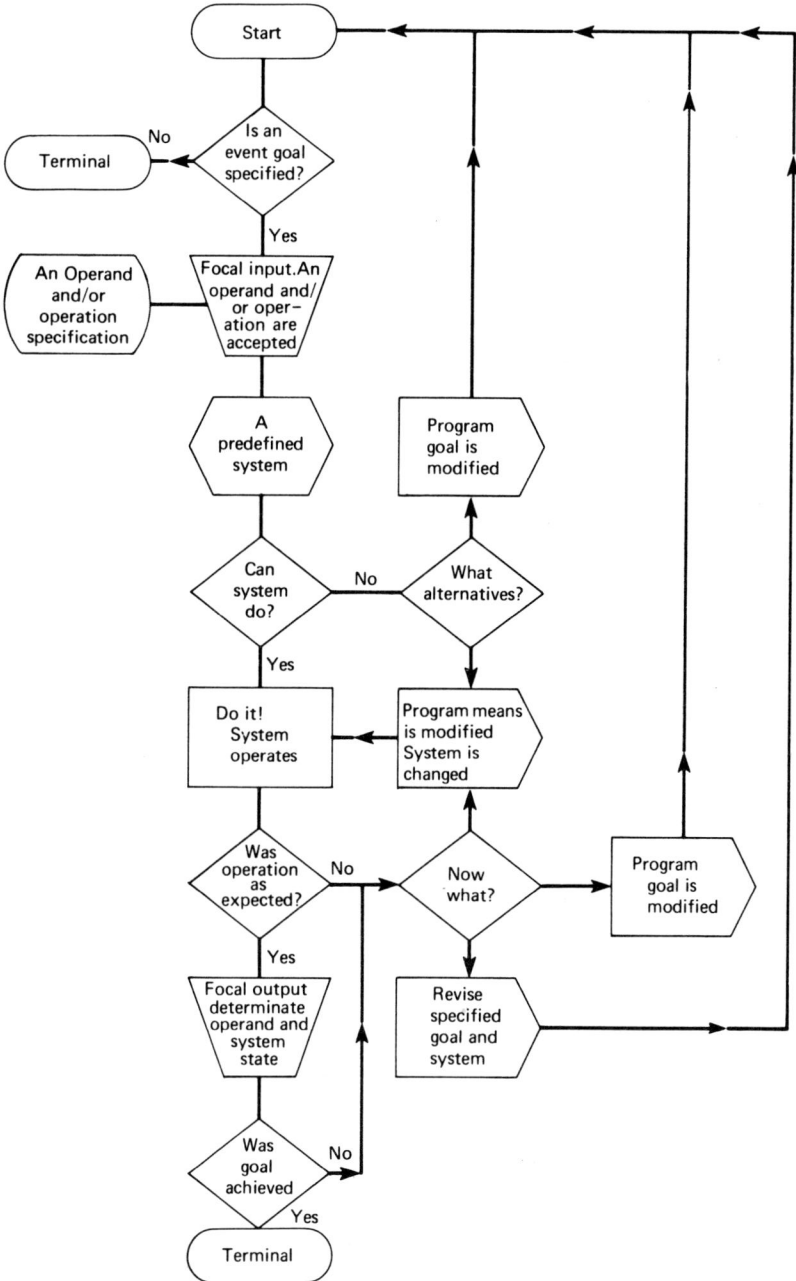

FIG. 4. Flow diagram of a transformation or event in an educational system.

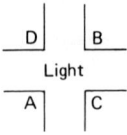

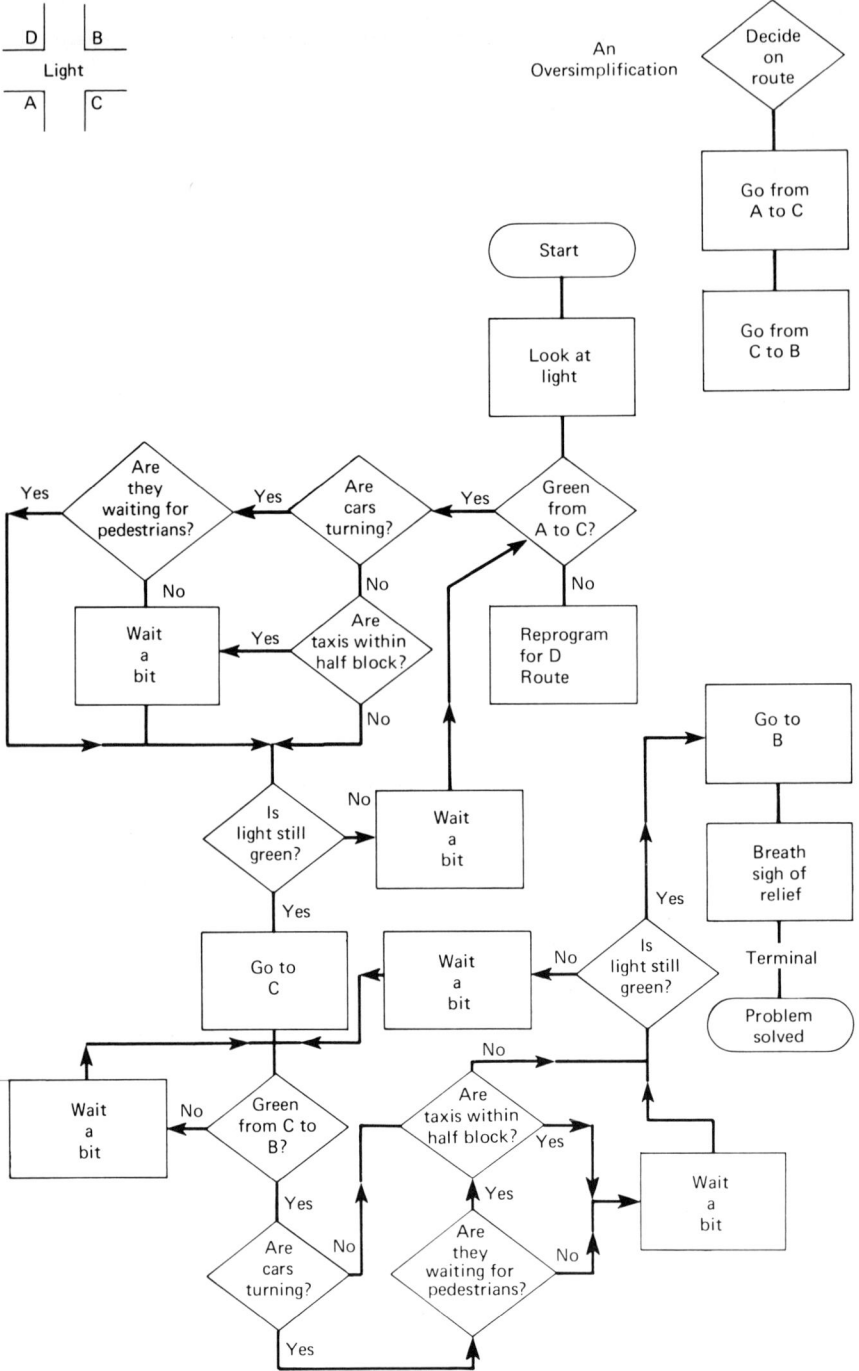

FIG. 5. Flow diagram: How to cross an intersection in New York.

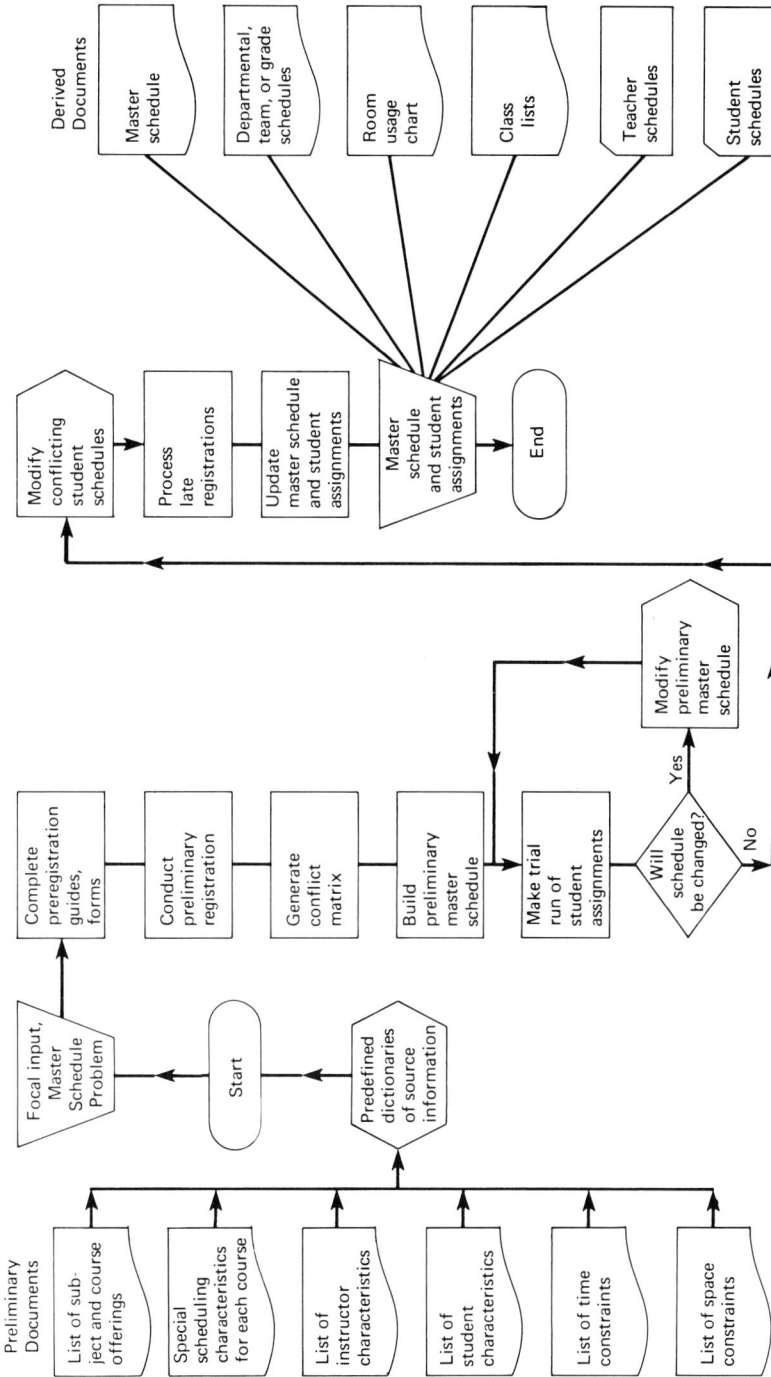

Fig 6. Systems flow chart of master schedule process.

tional flow depicted in this chart contains a feedback loop which permits a heuristic or search operation or series of trials which may be successful or may require some sort of program modification.

The final figure in the series of system flow charts, (Fig. 7), illustrates in a very simple manner the complex educational process of curriculum planning and development. It is shown here to demonstrate the need to plan for the articulation of many resources (even multiples of human resources) in most educational planning operations. It also demonstrates the several major milestones or subprograms involved in progressing from the initial planning stages of curriculum development to the final stages of general program implementation. Interested educators are referred to the abundant literature on flowcharting and scheduling for further information regarding the purposes, rules of thumb, and advantages of these planning techniques.

At this point the reader also is encouraged to discover the utility of system flowcharting for himself. An experience of this kind will frequently reveal to a beginning planner the human tendency to grossly oversimplify systematic descriptions and explanations of single production operations and events. He will soon learn the necessity of employing exact operational definitions and categorical specifications of system input elements and output responses. With practice the system flowchartist will also learn the importance of carefully predetermining decision points for precise monitoring of series or networks of events.

PERT/CPM (PROGRAM EVALUATION AND REVIEW TECHNIQUE/CRITICAL PATH METHOD)

Among the techniques employed by system analysts, operations researchers, and organizational managers are procedures for preplanning and prescheduling initial or one-of-a-kind complex operations. The ability to plan, schedule, and monitor economically costly or complex operations such as the construction of a new school or the development of a new curriculum is enhanced by such methods. Careful planning and scheduling reduces input omissions, permits better coordination of special projects with on-going operations, and reduces the probability of wasted resources, the tying up of capital, and the delay of production schedules.

The literature on such systems of cost-benefit planning and control is currently inundating journals devoted to business administration. In fact, the diversity of management scheduling techniques or systems is something of a problem for the federal government which has to coordinate and monitor its budget with information derived and reported via these

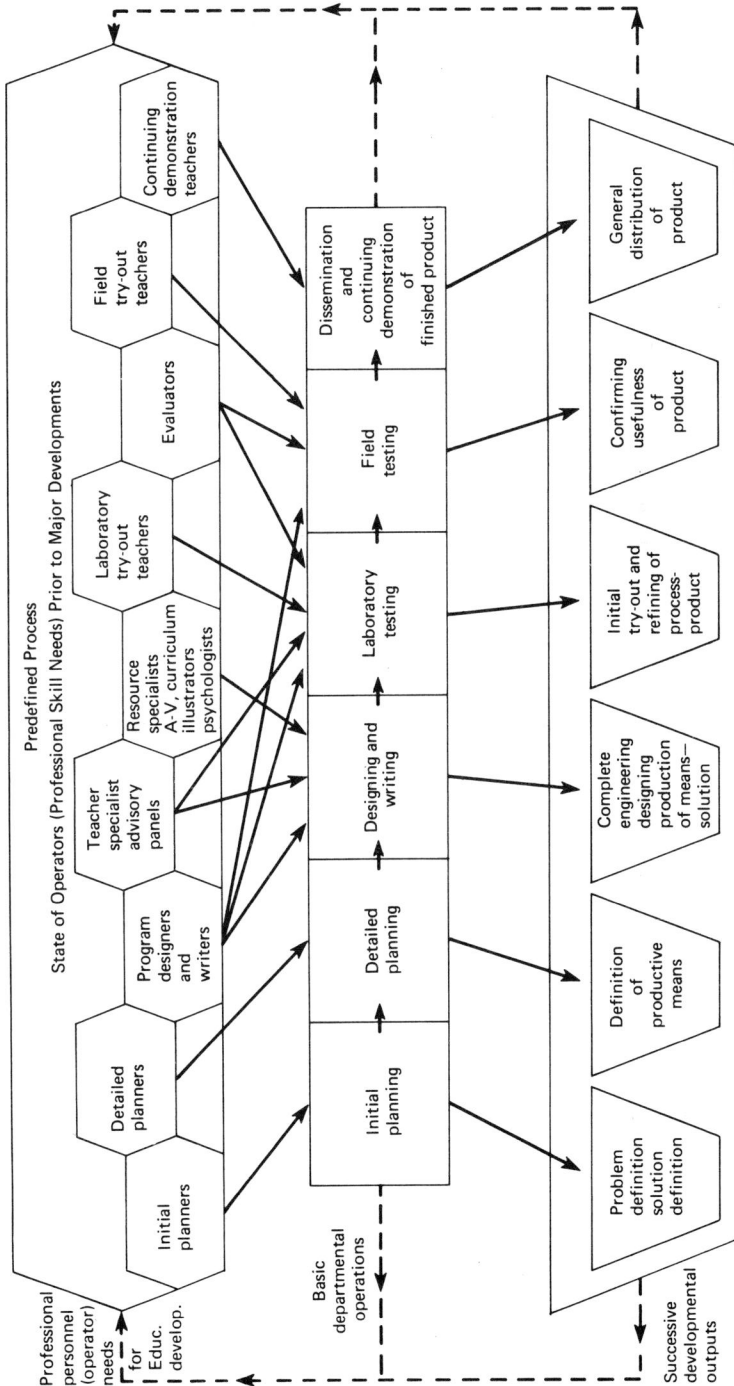

Fɪɢ. 7. Systematic educational development.

diverse means. There are currently more than ninety scheduling techniques possessing their own well-known acronyms, all of which are employed for improving the effectiveness and efficiency of planning complex production processes.

Because these methodologies are being employed more and more frequently by professional educators for major project planning, an introductory explanation of the most widely known process is presented here. Selected for exemplary presentation and interpretation is the PERT/CPM process. PERT was originally developed by the Navy as a process for controlling the research and development of the Polaris missile. CPM was first used by DuPont and Remington Rand in the planning and monitoring of complex construction processes. Essentially the same process, PERT/CPM is a pictorial plan, schedule, or budget designed to guide planners in initiating and controlling initial or one-of-a-kind complex system operations or networks. (Kaimann, pp. 43–57) Some processes involving as few as 200 man-hours of labor have been PERTed, however.

The essential concepts comprising the PERT/CPM planning system include the following:

1. A major project or organization consists of many individual elements and activities. Some of these, by their nature, must be introduced in a serial and incremental order. Other elements or operations may be introduced simultaneously, independently, or alternately. The coincidence of significant inputs in space-time-function define major event points where previously independent elements intersect and interconnect.
2. Various significant inputs and activities may be linked together and displayed on a flowchart or network. The resulting structure or model forms a visual display of the system.
3. Time coordinates, together with spatial and operational coordinates, are useful in describing the system in advance of its operation and in controlling its efficiency during operation.
4. The longest time path required within the serially ordered network of events determines the time required to complete the project. This path is called the *critical path.*
5. All other time paths of events throughout the network have some time *slack* relative to the critical path.
6. Efficiency in the system is often created by correctly assessing the critical path, and, if possible, ascertaining ways of reducing it. Making maximum use of elements and resources not in the critical path is another way of inducing efficiency in the planning system.

The process of developing a PERT/CPM chart begins with the identification of all necessary and sufficient elements and operations in the system network. Total enumeration of these elements includes the initial identification of the system's output goals and all procedures and processes necessary for their attainment. All resources and services not presently under full control of the organizational planner must be accounted for in the scheduling procedure. A time schedule of each input/output element and event within the total network must be developed as follows:

1. A completed PERT/CPM network model depicts the plan to be used and the time required for the project. The organismic dimensions of the project are specified. Its primary goals (feedforward) are established; the necessary subprocess objectives are identified. The size and magnitude of the project is predetermined in the planning and scheduling phase.
2. The network model establishes the critical path (the longest required time path), the pattern of exact interrelationships or interconnections among system elements in event milestones, and the technical requirements for completing each event in the primary and parallel activity sequences or paths.
3. The specification of events and their technical requirements identify the boundaries or limits of money, manpower, and time resources needed for each event and for the project as a whole. How much money and when it will be needed or will be available must be established in advance. Manpower specifications indicate skills of individuals or organizations which articulate into the network. Approximate working deadlines are predetermined by estimation and/or calculation.

This quality and quantity of information aids in planning and scheduling. It facilitates coordination and communication among working units. With sufficient information and organization an ordered series of predefined points for monitoring the system's operations through current feedback can be predetermined. Such cybernetic control systems can identify potential problems early.

Thorough preliminary planning and scheduling may permit rather complete simulation or predetermination of the effects of alternative decisions under consideration, providing an opportunity for testing and comparing their benefits and costs and noting how they affect the total program in the network. There are numerous mathematical and statistical techniques used by operations researchers for estimating and calculating cost estimates and useful levels of schedule completion probabilities that are applied in PERT/CPM processes.

PERT/CPM Network Symbols

The flowcharting of a PERT/CPM network has a number of established conventions that are generally followed. (Cook, pp. 93–94). The fundamental symbols are those which identify different levels of activities or events. The basic network symbols used in PERT/CPM are:

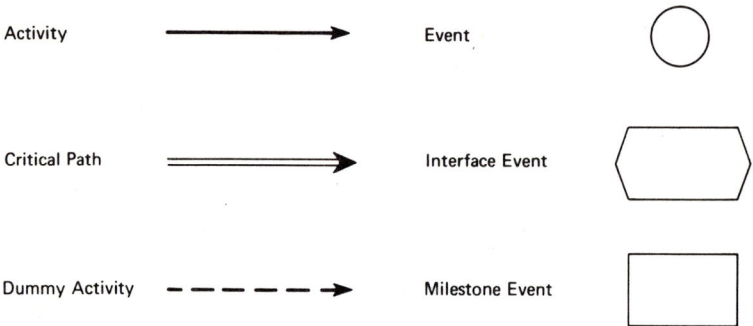

Activity	———————▶	Event	◯
Critical Path	═══════▶	Interface Event	⬡
Dummy Activity	– – – – ▶	Milestone Event	▭

Activities are time-consuming elements such as minor jobs, tasks, procurement cycles, or waiting periods. They represent the time (sometimes resources) necessary to progress from one event to the next. The common activity symbol employed is the flow or directional line, usually overlaid with an arrow direction indicator. An activity or process cannot begin until all previous prerequisite activites have been completed and all resources secured. Resource allocations include manpower, materials, equipment, and facilities.

An *event* in PERT/CPM is a boundary point or terminal which indicates the beginning or ending of an activity. It is usually indicated as an arbitrary or natural terminal point following a final activity or series of activities, a point which completes a subassembly (Cook, p. 94) The awarding of a contract or acceptance of a pilot model or plan is an event point. An event occurs and is identified within space-time function parameters or limits. Events in PERT/CPM are points of time, not durations or activites which are time-consuming. Events in PERT/CPM are similar to the terminal symbols employed in system flowcharting.

Event points are usually shown as circles or terminal symbols (elongated circles) and are often numbered or described. They may be defined by their service as preceding or predecessor events or as final or terminal events, wherein their definition indicates them as succeeding and/or successor events. It is important to remember that each activity or flow line must have a single predecessor and a single successor event.

In a PERT/CPM network the *interface event* occurs when one preceding flow or sequence of activities ends or intersects another; a point at which another phase or plan begins. The predefined process symbol used in system charting is often employed to indicate interface events. The operational symbol or rectangle is employed to indicate *milestone events,* key program or project accomplishments, in the network.

A broken line is used to symbolize a *dummy activity* in a network. Dummy activities do not involve work nor consume time or resources in the system under consideration. They merely indicate a necessary precedence or dependency between events.

A double activity line identifies the *critical path* in the network.

PERT/CPM Time Estimates

In PERT/CPM network diagraming, time estimates are assigned to all activity flow lines. In general three time estimates are made for each activity. These are: (a) an optimistic estimate, (b) the most likely estimate, and (c) a pessimistic estimate. (Kaimann, pp. 43–57) The optimistic estimate is considered as the minimum time necessary for action if everything fits. The pessimistic estimate is the longest time experienced in a similar action or an estimate accounting for bad luck. When multiple estimators are involved, the optimistic estimate sometimes represents the lowest estimate of 25 percent of the estimators and the highest estimate is the norm point at which 75 percent of the estimators believe the project can be completed by that date. The accuracy and effectiveness of the PERT/CPM process depends very much upon the completeness and correctness of the network logic (the pictured relationships) and upon the accuracy of the time estimates.

The three predetermined time estimates—the optimistic, expected, and pessimistic estimates—can be used to derive a mean time and variance for performance of an activity. They can be translated into frequency distributions and expressed in terms of statistical probabilities and distribution curves if desired.

When all of the events in a complex PERT/CPM network are defined and time estimates are made for all prerequisite activities, it is possible to determine the sum of all mean-time paths through the network. The largest of these sums is established as the calculated required time or critical path time essential for completion of the project. When the entire network is plotted and a start date or terminal date is known, the alternate date can be established. Prediction of the latest allowable start time derived by prior knowledge of a completion or contractual obligation date, is important for economically planning the project.

The Concept of Slack

If a PERT/CPM analyst has determined the latest allowable start time and the expected time of completion, it is possible for him to compute the slack in any of the activity paths within the network. *Slack* is defined as that time existing on every path in the network except the critical path which comprises the difference between the predicted activity time and the latest allowable time needed for completion. Slack may be expressed as a positive or negative value relationship. A negative value would indicate the necessity of beginning activity earlier or extending the terminal date of the project.

Paths with zero slack indicate sensitive or critical paths. Any deviation from the expected on these paths would force deviations and delays of the entire project. Paths with positive slack indicate areas where an activity may be allowed to slip at least to zero slack without causing delay or expense in the project. The economical way to plan a project is usually to eliminate all necessity of slippage by delaying starting activities and by not committing resources until absolutely necessary.

The Concept of Critical Path

The one or more paths with zero slack in the network are *critical paths*. A critical path is one which requires the longest time between the start date (the initial event terminal) and the final termination or completion date. When any delay occurs on this path the entire project is delayed.

A simple model may assist the reader in understanding the concepts of slack and critical path in a PERT/CPM network. See Fig. 8.

In this figure, event 10 represents the start of the project, and event 14 represents the final completion or milestone event. Three lines of independent activities, the paths from 10–11–14, 10–12–14 and 10–13–14 must be completed independently prior to their intersection at 14. In the project the critical path is route 10–12–14, a path consuming 18 days. Alternate or parallel path 10–11–14 is estimated to consume 12 days, indicating a slack of six days on this path. Path 10–13–14 consumes 14 days, permitting four days slack. Activities on the non-critical or slack paths can be delayed in start. Any delay on the critical path would extend the project. Reduced to its basic components, PERT/CPM is a network scheduling technique which requires analysis, linear and parallel scheduling, and cost estimation. The theory of network analysis is not new to scientists and engineers. They have employed this technique for years.

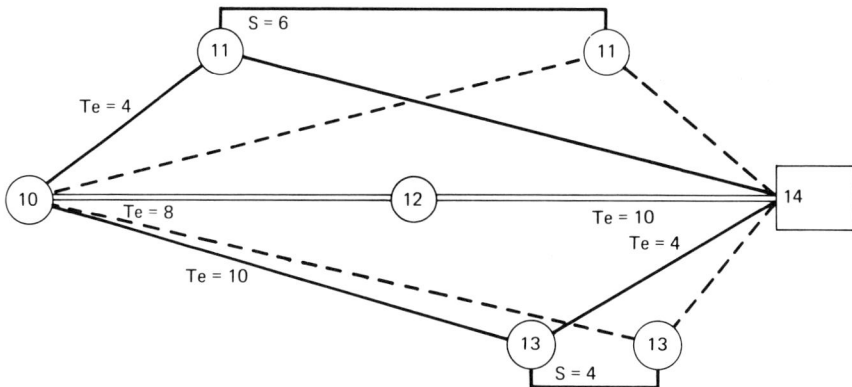

Fɪɢ 8. Slack and critical path.

Henry Laurence Gantt is generally credited with the development of the first charting and scheduling of production activities.

What is new about techniques such as PERT/CPM is the fact that they are designed to show the interdependencies and interconnections among both simultaneous or independent and incremental or dependent activities in complex operations; they can also be employed to reduce costs or economize in the systems or networks they represent. This was not done in earlier techniques.

Once the "links" between various essential operations in a complex network are clearly indicated three basic questions must be asked about each operation in a PERT/CPM network:

1. What inputs must be assembled or completed before the operation can begin?
2. What can be undertaken concurrently?
3. What must immediately follow (is dependent upon) this activity?

This is the essence of the scheduling problem in PERT/CPM.

Another question must be considered after basic scheduling relationships among operations have been determined and charted: Is there a way of optimizing the schedule to reduce either time or cost requirements by manipulating the elements, redefining or eliminating activities, reallocating resources, delaying inputs, or speeding completion of the project? Determining probabilities of a final project completion date is considered desirable also. A probability below 25 percent is considered too loose to be a useful predictor, and a probability greater than 75 percent is perhaps unrealistic as a predictor of pioneering or one-of-a-kind operations. It should be noted, however, that careful scheduling of all suboperations in a network reduces many uncertainties in predicting scheduling of the network as a whole. The result of applying PERT/CPM techniques is a much smaller maze and margin of error to be searched or estimated.

Although PERT/CPM provides an optimum schedule or network of essential activities in a system, it is based initially on an assumption of the availability of unlimited input resources, including time, for completion of the operation. Only through the use of further heuristic programing which requires the use of basic rules of thumb for economizing and scheduling can initial operations of this kind be made more efficient or economical. Some of the essential rules of thumb employed in PERT/CPM scheduling procedures include:

1. Break up the total time allotments into smaller scheduling components or increments. Initially schedule all jobs possible for the first time period, etc.
2. When several tasks compete for the same input resources, give preference to the operations with the least slack time.
3. Reschedule non-critical tasks (those not on the critical path) if possible, in such a way as to free resources for accelerating the critical path or reducing the time in which noncritical resources are tied up.

In summary, PERT/CPM analysis produces a visual model which clearly depicts a number of interrelated essential network activities, an estimation or calculation of the network's time dimension requirements, the designation of a critical path, and the heuristic programing of these components so that the final plan optimizes output accomplishments and input costs.

The simulated PERT/CPM network is a valuable administrative instrument or tool which requires a minimum amount of technical knowledge and programing skill. It relies primarily on the sincere efforts of an administrator to develop an accurate description of technologically workable objectives and their linear and parallel sequences of preceding activities or inputs. There is no inherent reason why this technique cannot be applied to almost any operation involving moderately complex or complex network operations.

PERT/CPM reflects most of the benefits associated with other system procedures. It permits consideration of alternatives prior to commitment of material resources. It facilitates prediction and estimation of costs, and it provides an orderly measuring device with predefined critical points for monitoring project operations. Its values are dual. Its initial value lies in the improved planning experience developed in the process itself. Its output or ultimate value lies in increased administrative efficiency brought about by systematic resource control or planning.

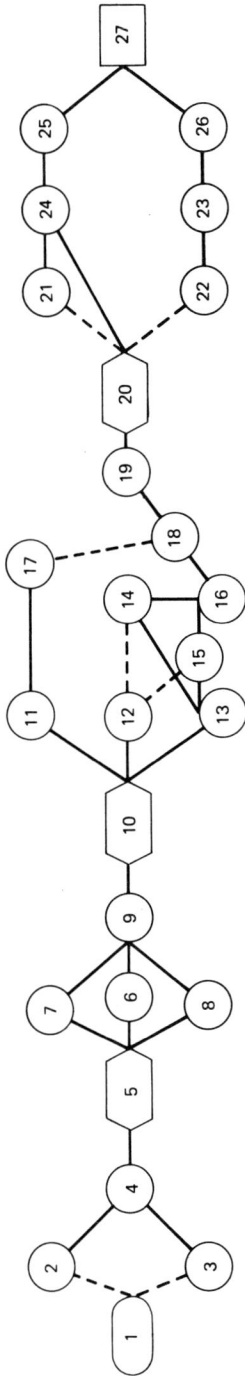

Task Identification List:

1. Start Project
2. Start Enrollment Projection
3. Start Educational Program Projection
4. Start Educational Needs Summary
5. Complete Needs Summary
6. Start Plant Evaluation
7. Start Financial Evaluation
8. Start Staff Evaluation
9. Complete Facility-Program-Staff Evaluations
10. Complete Long Range Plan
11. Start Building Support Plan
12. Start Architectural Arrangements
13. Start Specifications Guide
14. Start Quantitative Dimensions
15. Start Qualitative Dimensions
16. Complete Specifications Guide
17. Complete Building Support Plan
18. Complete Site Selection
19. Complete Architectural Arrangements
20. Begin Construction
21. Start Equipment Selection
22. Start Staff Selection
23. Start Staff Inservice
24. Complete Construction
25. Complete Administrative Transfer
26. Complete Staff-Student Orientation
27. Project Completion

FIG. 9. Preliminary PERT/CPM network for school building project.

Among the many applications of PERT/CPM one finds research and development schedules, budget schedules, construction plans, schedules for developing computer programs and installations, bid and proposal timetables, distribution planning, planning and scheduling of maintenance operations, and cost reduction reviews.

School administrators have employed PERT/CPM or similar techniques to plan campuses and buildings; install data processing systems; design new curriculums; prepare school budgets; schedule research and development projects; and plan testing, evaluation, and review procedures.

An illustration of a PERT/CPM network is presented in Fig. 9. This figure represents a preliminary network model or plan for a school building construction project. The network indicates the order of flow of twenty-seven requisite tasks common to most building projects, indicating their general linear or parallel (independent) order. A more sophisticated PERT/CPM network would show one or more time estimates of each activity and indicate the critical path through the network.

10

MARS—A Model Analysis and Redesign System

In this chapter a model or method broadly representative of basic system analysis and redesign rules and procedures is developed and explained. The MARS model is macrocosmic and interdisciplinary in its nature, applicable for analyzing any system or universe. It is substantially an ordered elaboration of the scientific method, which measures a limited number of variables in a controlled experiment or universe. System procedures must fulfill the additional requirement of integrating experimental events into standardized routine production operations in complex organizations. The MARS model begins and ends with a macrosystem or total system, but it does accommodate component or subsystem analysis, experimentation, modification, and planned reintegration into the macrosystem.

This model system analysis procedure is the foundation for a number of other system processes which apply its basic rules or logic and the sequential order of its operations toward the modeling of particular kinds of systems or programs. Some special adaptations of system analysis methodology in whole or in part are employed in PERT or critical path planning, operations research, simulation and gaming, programed budgeting, and learning unit planning and programing.

While considering this chapter, the reader is encouraged to note that MARS models primarily an iterative or recycling process which is updated periodically, particularly when significant changes occur in input resources or output goals. Any system procedure is heavily dependent upon this periodic review of data in its complex feedback network and the quality of feedback which the planning, operating, and evaluating processes are designed to stimulate.

In system analysis and redesign strategies, an analyst usually identifies from ten to twenty major steps or ordered subcycles which are presented in sequence and which can be recycled entirely or in part. Plans or models such as MARS can be applied in the analysis of extremely complex systems or in the analysis and restructuring of single events or simple subsystem components. In macroanalysis, component or subsystem analysis is generally treated as a subroutine or nested loop process which requires that any modification in a subsystem be evaluated independently and then integrated and reevaluated as it affects the performance of the system as an organic whole.

The steps in the flow diagram representing the MARS system will be explained more fully in the paragraphs which follow. See Fig. 10.

STEP ONE—FORMULATE THE PROBLEM

The first step in any systematic planning procedure is to formulate in a single precise sentence a statement which defines the focal problem or system involved. This statement will stipulate what the problem is, why it has been selected, and what comprises the system or program involved in its resolution.

STEP TWO—DETERMINE THE SYSTEM'S MAJOR MISSIONS, PROGRAMS, OR OBJECTIVES

The second step in the planning procedure is to identify one or more missions, programs, or major objectives which the system and plan will undertake to accomplish. Each major objective is explicated in a single sentence, employing behavioral, operational, or performance terms. The objectives will contain general information regarding all of the following conditions or variables:

What is to be done?
Where is it to be done?
When is it to be done?
How do we know if the program is successful?

If the program objectives are well stated and complete they will establish useful parameters for the remainder of the planning procedure and will serve as validating criteria for making later judgments of the system's suitability and effectiveness.

During the development of the system's objectives, an independent analysis of the functional constraints of the operating system or "doing

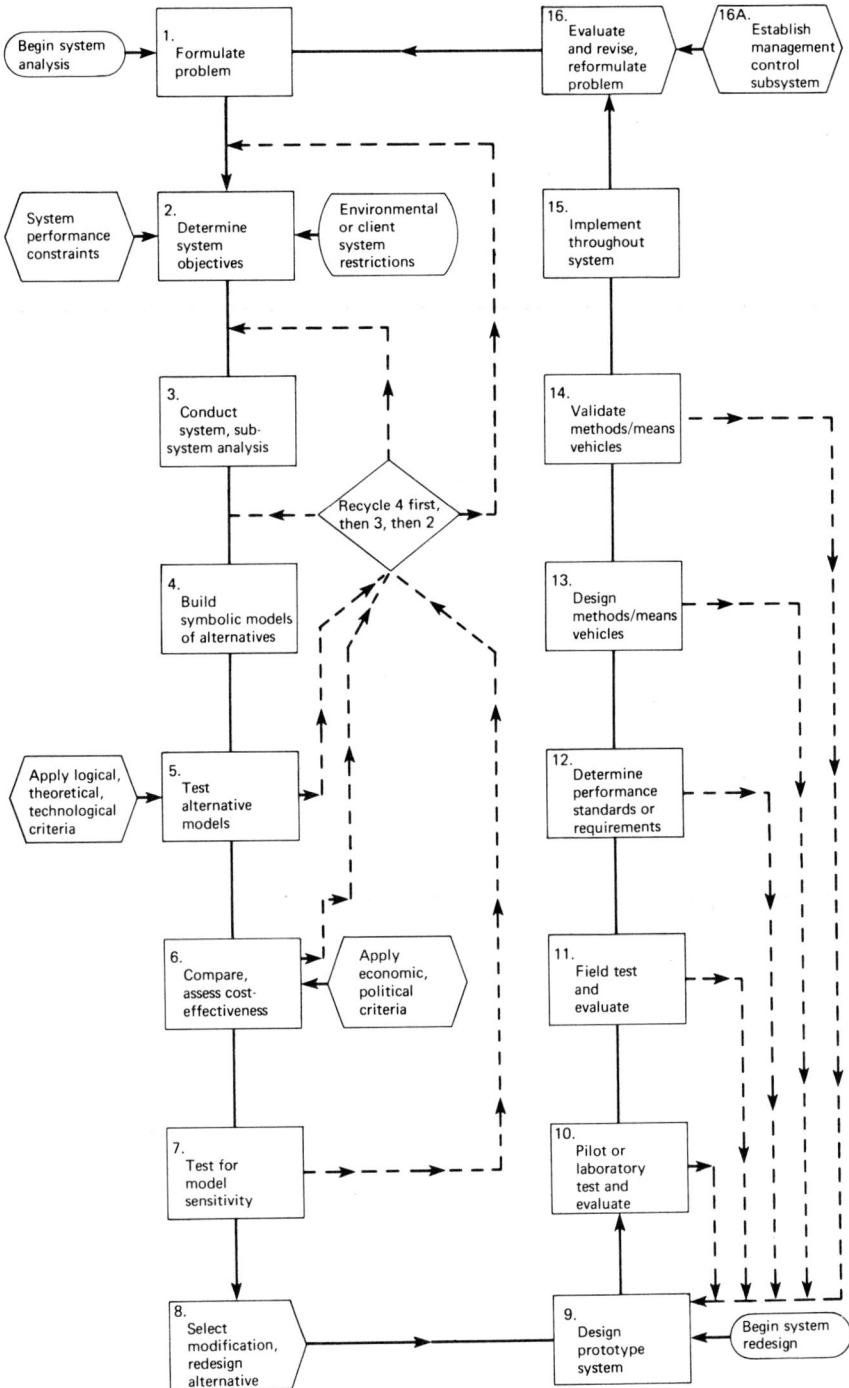

Fig. 10. MARS. A model analysis and redesign system.

agency" is undertaken. Among the internally controllable but limited resources or constraints inventoried in this step of the analysis are:

The human capital involved—numbers, skills, etc.
Buildings, land, and other physical space resources
Technology and materials
Financial resources and support
The system's time dimensions

In addition to the functional constraints or controllable resource limits influencing the operations of the system, there are a number of environmental restrictions which also shape and determine to a considerable degree the direction and rate of systems planning and implementation. A separate inventory of these uncontrolled variables is made at this time. The restrictions set by a client or customer system and/or other environmental system factors frequently establish limits upon:

Performance expectations, including the legitimate use of authority
and power
Time
Money—its legitimate or legal restrictions

Relevant laws and regulations of higher governmental agencies or supra-systems generally are included among the uncontrolled external restrictions upon a plan.

In combination, the specified programs or missions of the system, the controllable constraints upon the system's resources, and the uncontrollable restrictions set by the environment or client system pretty well determine the boundaries or ground rules for all subsequent steps in the analysis procedure.

STEP THREE—CONDUCT AN ANALYSIS OF THE SYSTEM AND SUBSYSTEMS

The third step in the analysis procedure is one of several extensive data gathering and analysis phases. In this step the program objectives become the system's specifications, establishing a tentative hierarchy, queue, or linear order of values for determining subsequent operations. They are predefined coded images of the output or product values to be attained by means of the system's operations.

At this time the analyst makes a thorough inventory of all resources that conceivably might be applied in the accomplishment of each stated program or mission. All resources or input systems currently or normally under the control of the macrosystem are included in this enumeration. In addition, the analyst will identify resources outside the system which could be obtained and which might prove to be preferred means-methods for accomplishing operations and tasks required in the plan. Selection of alternative resources is usually most productive when it begins with resources applicable to the following list of problem or program solutions, listed in an ascending order of constraint and restriction:

1. The theoretically ideal solution (no restrictions or constraints)
2. The ultimate possible solution (minimum reality constraints and limits)
3. The technologically workable solution
4. The practical and likely acceptable solution
5. The optimum solution

In this stage of analysis, planners should encourage a rich input of suggestions which are free from restrictions beyond possibly the requirement that they should be technologically workable.

When a rich reservoir of suggested program elements and operations have been identified, they should be organized into proper categories within the systems hierarchical order, according to function. Subclasses of macrosystem missions, programs, or major objectives usually can be decomposed into:

1. Major functions, problem subcategories, and related resources
2. Specific operational tasks or program elements within functions
3. Specific means-methods-materials needed for program task completion

An expert system analyst identifies several plausible alternative means-methods inputs for each task, alternative tasks for each function, etc., even though the alternative inputs may not presently exist in the system.

It is at this stage that free divergent thinking plus subsequent induction control (down to the workability level) pay off, especially if they are based upon rich experience and training. Occasionally intuition and creative effort may make the search for alternatives even more productive.

STEP FOUR—BUILD SYMBOLIC MODELS OF ALTERNATIVE PLANS

Since system analysis is of no value where it does not provide alternatives to present output objectives and input resources and operations, both inputs and outputs must be provided with options if the maximizing of optimizing choices is to be made possible and probable. The generation of alternatives is not a particularly difficult procedure, however. Means-methods and goals singly or in combination may be varied in the numerous ways listed below:

Inputs or outputs may be initially added or completely eliminated.

The degree of emphasis upon outputs or inputs in an operation may be changed.

Additions and subtractions of subtasks may occur.

Substitutions of inputs and outputs are possible.

Different combinations or groupings of input resources can be made.

Any of the variables in the system can be sequentially reordered.

New materials, technology, environmental conditions can be considered.

Personnel resources, activities, roles, and groupings can be altered.

After all alternatives to present system and subsystem inputs and outputs have been ennumerated, at least one model or plan incorporating and synthesizing new elements and operations should be developed in a detailed and organized manner. The existing or conventional system usually serves as the basic model for comparison, because its operations generally have been thoroughly tested and its component parts usually provide a rather complete catalog of necessary elements and operations. Frequently, desirable alternative plans may include adaptations of systems in extensive use elsewhere, as these sources possess similar advantages.

In this stage of the analysis it is necessary to determine an order or sequence of input resources and events that will likely produce the desired output states of first, task objectives, and finally, program objectives. The planning at this stage also takes cognizance of the original problem statement and the statements of the system's programs or missions. It also considers the system's constraints and environmental restrictions. The proposed plan should be generally congruent and consistent with all of these parameters.

The number of alternative models or plans developed in this step of the analysis, the degree of originality, and the depth of detail in each plan are somewhat dependent upon the resources available to conduct the analysis and redesign process. Extensive change in complex systems is generally very expensive. The reader should keep in mind also that productive search is generally accomplished by controlled rather than uncontrolled induction, by varying only a few inputs or outputs rather than by attempting a total system redesign. Scientists point out that incremental change is usually more effective than major reorganization. Experts often say that productive search is simple-minded.

Yet there are times when major reorganization or macrosystematic change proves to be a better alternative than one or more subsystem changes. If a major change is undertaken, it is to be expected that a significant allocation of resources for planning and implementation is needed. Usually such changes require both extensive retraining of personnel and additional technological and material resources and facilities. In many cases present systems of operations must continue simultaneously with the phasing in of significantly new systems and procedures.

In the planning of complex system analysis and redesign in educational organizations or other social organizations there are almost always multiple programs or mission objectives to consider. In such planning the primary program is likely to be that which accomplishes the organization's basic production or service objective(s). However, system analysts may give major or even focal attention to planning such auxiliary systems as the budget system; policy, records, and reports system; personnel system; management control and evaluation system; data processing system; transportation system; and inventory system.

STEP FIVE—TEST THE ALTERNATIVE MODELS OR PLANS

The fifth step—the testing of alternative models; the sixth—assessing and comparing their cost-effectiveness; and the seventh—testing the models for sensitivity—are also extensive data gathering, data analysis, and data reorganization operations. If a computer simulation of an operating system is undertaken, it is probably carried out here. Research of the literature and experimentation related to the effectiveness of little known subsystem components are completed in these steps. Original technological inventions are planned and tested at this time.

During the initial stage of the testing of alternative models or plans, the analyst develops an operational planning procedure or schedule which guarantees that the maximum information and personnel resources in the system are indeed available at the right time and in the right place to assist in productive decision-making. This usually means that all of the experts in the system participate productively at some time in Step Five. If the analysis and redesign process undertaken involves a system or program of great complexity, cruciality, and value or expense, considerable research, experimentation, and simulation are usually necessary to provide the requisite information needed in these three steps.

The first set of tests imposed upon the alternative models or plans should be the tests of logic. Logical analysis includes the consideration of these questions:

Are the terms employed in defining program inputs and outputs designative and meaningful?

Do the detailed statements of component operations seem to be consistent with the broad goals in the original problem statement?

Have the system's constraints and environmental restrictions been fully considered?

Are all components considered in each alternative master plan?

Have the analysts considered all of the critical implications and inferences in the proposed plans and programs?

The second set of tests in Step Five direct the analysis to a consideration of the proposed programs within the framework of current theory and research. Experts in all relevant disciplines give assurance that the plans are consistent with current disciplinary theory and research. They carefully search for inconsistencies between current findings of research and the theoretical inferences and assumptions contained in the several plans.

A third set of tests applied in Step Five are the tests of technological feasibility. Substantive answers to the following questions are confirmed in some systematic way:

Are the human skills, machines, instruments, facilities, materials, soft-ware, etc. available for implementing the plan?

Can they be obtained within the required time?

Is the plan technologically impossible?

Do prototype models of some elements need to be tested extensively before certain subproblems can be resolved?

STEP SIX—ASSESS AND COMPARE THE COST-EFFECTIVENESS OF ALTERNATE MODELS

Alternative plans or programs may differ in output expectations and in input resources and expectations both qualitatively and quantitatively. Step Six is the stage of development in systematic planning where these differences are carefully identified and meaningfully compared and evaluated. The procedures for evaluating expected benefits and costs of major programs require the involvement of both technical and policy-making expertise. They will probably require group processes in all policy-making and review activities plus considerable individual technical research and data manipulation as well.

It is most necessary at this time to identify clearly the economic benefits and costs estimated to accompany each program or objective. If possible the benefit-costs of each program element or means-method input and output should be independently assessed. A comparison of the alternate plans or systems as organic wholes is of course essential also. Certainly the comparison of alternate models will show the need for major capital inputs into the system. Expert calculation of relative costs and benefits will take into consideration both the short-term advantages and costs of program elements and total operations and their long-term benefits and costs. Major redesign of systems seldom occurs if only short-term economic advantages are considered.

The ideal economic cost-benefit analysis of alternative plans requires the development of a valid and reliable indexing system which directly and fully compares the plans in all of their critical aspects and indicates in a single summarizing index figure a relative order of value. In operations research an economic or monetary index is often used to produce an optimum decision. Operations research techniques assume that their calculations are based upon such complete information, although operations researchers use techniques of estimation as well as calculation in their processes. When based on considerable prior knowledge and experience, operations research procedures are particularly useful in determining estimates of how much to spend or produce in certain programs or what optimum input level is needed to produce the maximum profit. Educators should note that the optimum input level of a system is seldom the maximum level, and that the relationship between input level and output level is seldom direct or linear.

In contrast to operations research technicians, system analysts are often concerned with programs where the number of assumptions and estimates required by a mathematical indexing system are so gross and numerous that the calculation procedure contributes nothing toward sharpening the real differences among the alternatives.

In the use of calculation procedures involving economic indices, it is assumed that the indices are truly representative of all benefits and costs involved in the alternate programs. In reality an economic index is considered excellent if it represents reliably and accurately only the measurable monetary costs and benefits. In complex social organizations with non-monetary benefits and costs, it is almost impossible to calculate decisions which actually represent both monetary and non-monetary values of inputs and outputs.

There are, however, situations in social organizations where the calculation of decisions among alternative programs is heavily dependent upon the constraints of a monetary nature. In such situations an analyst employs a number of standard procedures or rules of thumb for making economizing decisions in his planning. He may consider:

1. Elimination of the need and therefore the function (subsystem or total system).
2. Specifying as few low-cost inputs as possible.
3. Specifying as few low-cost outputs as possible.
4. Utilizing personnel skills one hundred percent of the time, and at the level of their maximum capability and applicability.
5. Utilizing other resources efficiently.
6. Designing systems for regular or normal conditions before incorporating exceptions.
7. Automating data handling and, where possible, production operations.
8. Building into the system adaptive and flexible feedback controls.

While monetary indices are employed by operations researchers to sharpen or calculate decisions among alternative plans involving complex programs, it is difficult and dangerous to ignore political or social values in favor of economic values only. Some experts suggest that political values are in fact basic and prior to all others in their importance and function. Certainly the political benefits and costs of alternate models should be considered in making decisions regarding important social missions such as education. An educatonal administrator knows that political intuition and information is frequently as important as economic intuition and information in planning and fully implementing changes in educational institutions.

Astute social-political assessment of the political impact of alternate plans can be aided by systematic methods. Techniques of polling, personal-social opinion gathering, and methods of market research can improve the reliability of political assessments. However, a system analyst should be aware that the making of social-political assessments involves an extremely high degree of uncertainty. In addition to statistical or chance uncertainty in the natural and technological systems involved, there is additional uncertainty in the human communications systems and subsystems. This includes both strategic uncertainty, uncertainty due either to error in planning or to counterstrategy, and tactical uncertainty, the uncertainty caused by human caprice.

If a cost-benefit assessment of alternative programs includes both a social-political assessment and an economic assessment, these procedures will probably be carried out independently and reported separately. A careful analyst will have to weigh both the political importance attributed to economic factors at any given moment and any independent political momentum or climate. Only when the analyst believes that his cost-benefit methodology has independently calculated the characteristics of alternate plans in respect to both economic and political factors will he attempt to index them in such a way that they can be meaningfully combined.

If a system analyst believes that such a procedure is both possible and desirable, his final step in the cost-benefit assessment and comparison is to combine all costs and benefits—technological, economic, and political—into a single concluding judgment, calculation, or statement. An ideal plan will, in a single mathematical indexing system, permit calculations which will minimize technological, economic, and political costs and maximize their benefits.

STEP SEVEN—TEST FOR MODEL SENSITIVITY

In the testing of models or plans of extremely complex programs it is seldom possible to actually calculate meaningful and reliable indices which indicate a precise comparison or an optimum choice among alternatives. When a system analyst believes that qualitative uncertainties attributable to presently unknown but significant predispositions and subsequent human counterchoices or counterstrategies are greater than statistical variation (chance variations in the real world or in his predictive models), he is confronted with a problem of great uncertainty. Human judgment must transform inconclusive information into wise decisions in such circumstances.

There are a number of strategies employed by analysts for sharpening the perception of differences among alternative plans in such instances and making later judgments less intuitive and more surprise-free or systematic. Procedures employed to facilitate the sharpening of decisions in extremely complex, ill-defined, and prescientific decision situations are called sensitivity analysis processes.

Sensitivity analysis is designed to redirect an analyst's attention deliberately by means of an explicit treatment of uncertainties. It requires him to rank alternate plans under a host of assumptions. In sensitivity analysis the planner deliberately departs from the expected, normative, or predictive input or output value ranges and uses instead several arbitrary variations or estimates of these values—for example, high, medium, and low quantities or rates, and good, better, and best qualities, conditions, etc. He may vary either quantitative or qualitative dimensions in this way.

Contingency Analysis

One type of sensitivity analysis, *contingency analysis*, attempts to see how sensitive the rankings among alternate plans are to variations in either program inputs or outputs—variations such as changes in the criteria of evaluation, major changes in the environment, and significant changes in the controllable or internal resources of the system—for example, staff changes. Often three levels of input quantity or rate variations or output expectations (high, medium, and low) are hypothesized and the systems are compared according to their ability to accommodate these variations. One system may be incapable of effectively or economically adjusting to such variations. Another may become less effective or economical when inputs are increased or decreased. The better system may be second best when contingencies are changed. Through contingency analysis planners seek to identify the all-around best system for efficiently performing within the probable range of contingencies.

Contingency analysis may include an analysis of plans with a view of testing their sensitivity to changes in the time phasing of their expected rate of implementation or production, determining which model adjusts best to an acceleration or deceleration.

Another type of contingency analysis is made possible by deliberately rescaling or reranking the order of emphasis placed in listing the program objectives or in establishing criteria for the planning procedure. It questions how the plans will be perceived if economy is deemphasized and performance is given more emphasis, or vice versa. It inquires into what happens if political benefits or costs are given more or less value in comparison with values assigned to technological or economic factors. It determines if there are any new theoretical or value criteria which can be introduced to give the planners a richer frame of reference for deciding among the alternatives under consideration.

A Fortiori Analysis

Another type of sensitivity analysis is called *a fortiori analysis*. Through the use of a fortiori analysis, planners seek to control the human bias existing in many planning or analysis processes. In a fortiori analysis planners bend over backwards to identify advantages and deemphasize disadvantages in plans or models they intuitively dislike or rank low. If, after doing this, the previously preferred plan still looks good, it may indeed be the best.

Break-even Analysis

Still another type of sensitivity analysis is *break-even analysis*. Break-even analysis also tests the sensitivity of judgments among alternate plans. In break-even analysis the analyst changes the rank order preference of plans to make them appear even. Then he considers differences among the assumptions used to determine both the original ordering and the reordering. This procedure often reveals prior biases which may have been overly optimistic or overly pessimistic.

STEP EIGHT—SELECT ONE PLAN, MODIFICATION, OR REDESIGN ALTERNATIVE

Step Eight in the MARS system of analysis is the culmination phase of the analysis procedure. In this step a final convergent decision or judgment is made which identifies one model or plan among the several under consideration as the best plan or program for subsequent implementation.

The first activity in Step Eight is to organize and codify a careful and meaningful presentation of the alternative plans. In this process their ranking may be indicated and documented. Justification for any ranking is presented in the form of technical evidence (logical, theoretical, and technological); evidence of economic and political feasibility and benefit-cost comparisons is presented. Any important findings derived from tests of sensitivity are presented also.

As a result of following the procedures outlined in the first eight steps of the MARS procedure, the analyst and/or his associate planners should be in a position to produce two final planning outputs. The first output of this series of procedures should be a well-organized planning document which serves as a valuable communications instrument and referent. The second output is the formalization of an irreversible decision reflecting the acceptance of a single preferred program or plan for implementation.

STEP NINE—DESIGN THE PROTOTYPE OR PILOT MODEL OR SYSTEM

The MARS system next begins the redesign stage, where abstract paper plans or symbolic models are transformed into real world operating programs and systems. This phase begins the synthesizing or designing of the material prototype or model system. It involves the systematic induction of material elements and operations in an orderly and incremental sequence. If prior analytic operations were well done, the design and implementation operations are relatively easy to accomplish. It is likely, however, that actual operations will assume a complexity never fully anticipated in the planning and analysis stages. The materialization, implementation, and evaluation of an operating program, even if previously well planned, generally require careful and heuristic guidance by its designers.

In the first step in the redesign and implementation phase of the MARS process, the analyst constructs a plan or schedule for the complete implementation process. Procedures such as PERT or Critical Path are applied at this time and updated, modified, and extended at some time in the future. A flow diagram of the order, sequence, and general time schedule of means-methods, tasks, and programs is the minimum plan that is acceptable. Included in this initial step of the redesign and implementation phase is the detailed designing and construction of a prototype or pilot model of the system under consideration or any of its previously untested components. An enumeration of all components needed in the pilot operation is necessary.

STEP TEN—PILOT OR LABORATORY TEST
THE PROTOTYPE MODEL AND EVALUATE

In Step Ten the prototype of all new or untested elements in the revised program is put into operation. All input materials needed in maintaining the prototype operation are assembled. Personnel, space, machines, materials, and money are allocated. The operations are set in motion and heuristic modifications are made when and where they are observed to be necessary.

When the pilot or prototype model seems to be accomplishing reliably what it was designed to accomplish, a thorough evaluation of it is in order. In addition to assessing its predetermined validity and functional reliability, it is studied for unanticipated outputs or side effects.

STEP ELEVEN—FIELD TEST AND EVALUATE
THE PROTOTYPE MODEL

Following the laboratory testing of all new components in the program, the prototype elements and operations are moved into the field for testing. In field testing, the designers thoroughly examine the system as it operates in an environment generally representative of all conditions and contingencies it will experience in normal production. Research on the human impact of the prototype and upon its "cosmetic" or "packaging" effect on worker or client motivation is undertaken.

After the system appears to be functioning effectively and efficiently in the field it is evaluated systematically once again. Major changes made during the field testing phase are thoroughly evaluated also. The system is validated and its unexpected effects controlled in a desirable or productive manner. An inventory of all of the materials, media-aids, and methods involved in field testing is made, and a log or record is kept of all critical incidents, both positive and negative in character.

STEP TWELVE—DETERMINE PERFORMANCE STANDARDS OR REQUIREMENTS FOR THE SYSTEM

From extensive and exhaustive field testing it is possible to determine with considerable accuracy the performance or output standards of the system or its subsystem components. These standards take cognizance of the specified system objectives and the performance standards expected from alternative systems. Although the system's capabilities may appear to be outstanding, the performance standards established for subsequent general production operations should be conservatively stated. They should be the product of valid and reliable field tests and should account for all contingencies experienced in the field testing process. The performance standards so stated can be used as initial criteria for evaluating the completed installation and its integration into the producing system.

Without much doubt the most critical component in the setting of performance standards for a program or operation is the human factor. A thorough analysis is made of all human skills required for the effective operation of the system at this time. The necessary information and applicative know-how required in actual operations are carefully determined. Human physical, mental, and emotional requirements are identified.

STEP THIRTEEN—DESIGN METHODS-MEANS VEHICLES

In this stage of the implementation process, specifications for facilities, equipment, and general support communications and logistics are established. Machines, materials, media, tools and aids, are purchased and/or produced. Personnel involved in the actual programed production are thoroughly trained in all of the tasks which they are expected to perform.

STEP FOURTEEN—VALIDATE THE METHODS-MEANS VEHICLES

The process of validating the methods-means vehicles includes all procedures which reduce uncertainties in the total system's operations that hinder the system-wide implementation of the prototype operation.

All necessary input resources are distributed and inventoried; the machines and technology are set up and tested in trial runs. Above all, the personnel involved in the program are field tested and evaluated.

STEP FIFTEEN—IMPLEMENT THE REDESIGNED MODEL THROUGHOUT THE SYSTEM

The new model or program can be said to be fully operational when it is in actual service, when every operation is undertaken, and when all components are in use and functioning well. At this time the system should be sufficiently accomplishing its predetermined programs and missions. If the program of the new system is considered critical, a standby unit of the old system is sometimes maintained temporarily for emergencies, but, if the new system is well-designed and properly installed, the old system can soon be abandoned and its properties sold or reassigned.

STEP SIXTEEN—EVALUATE AND REVISE THE OPERATING SYSTEM

A well-planned and organized procedure for evaluating and revising the model program is designed and fully implemented. Means for sensing and reporting on machine performances are established and operationalized. Personnel performances are monitored and evaluated regularly.

Changes in the original management control are made to provide for supervisory assistance for all personnel in the new program. All other support systems of budget, finance and payroll, records and reports, inventory, communications, personnel, transportation, and health and safety are updated to facilitate the integration and coordination of the new program into the total system's operations.

SUMMARY

It is important to note that, within the MARS system or any other complete system analysis procedure, each step or task is composed of a number of related activities culminating with an evaluation or task recycling and revision before beginning the next operation in the linear order of activities. The entire model or system is of course recycled when the original problem is redefined, thereby changing basic objectives.

There are two additional important nested loops or complex recycling procedures inherent in system analysis. In resolving a difficult problem the planning or analysis subprogram may require search, trial and error, and recycling of Steps Two, Three, or Four through Seven. Likewise the design and implementation subprogram may require heuristic recycling of Steps Nine through Fourteen.

11

PPBS—Planning Programing Budgeting System

INTRODUCTION

One of the applications of system analysis which is receiving considerable attention from educators is PPBS—Planning Programing Budgeting System. In PPBS the methods and techniques of system analysis and operations research are applied to budgeting procedures with rather significant results. Budget instruments and planning procedures are becoming more comprehensive and complex, simultaneously more analytic and integrated. As stated previously, all system processes are sets of theoretical generalizations and standardized rules of procedure which were induced originally from primitive and unsystematic human experiences or trials, but, because these trials are subjected to continuous and incremental regulation, evaluation, and recording, they eventually attain an order and precision of meaning which converts them into highly workable and economical operational techniques.

A budget is a plan or model which describes, classifies, and allocates for accounting purposes monetary inputs into an organization or system. Traditionally, budgets have functioned as mere organized financial allocation systems. They have been employed generally as systems for allocating moneys for specific functions or work patterns, and their span of application is usually focal and of short duration, involving, for example, a special project or a single fiscal year.

Budget processes have changed with changes in social, technological, and economic conditions, however. The obvious and severe limits of economic resources available for planning and operating organizational systems at one time dominated the structuring of the budget process.

When monetary inputs were clearly limited in their adequacy and were perceived to be absolute operational constraints and restrictions on planning, financial resources were fully assessed and fully acquired before organizational leaders were allowed to make any allocation of funds to particular programs. Budgeting was and still is a process for planning the allocation of scarce or limited resources. In theory, economical budgeting must set tight priorities among all possible organizational needs and goals. The traditional or conservative budget may be called an input budget.

Encouraged by the accumulation of financial reserves in some organizations and by the acceptance of new economic principles of borrowing or investing in the future, the budgeting process began to change. Organizational administrators developed an increased interest in meeting other than urgent survival needs, in seeking the benefits of longer-range growth profits, in developing the profit potential of the leisure-time industry, and in allocating more time and other resources to organizational planning and evaluation functions.

The professionalization of management was another contributing factor to the improvement of the budget process. Managers became more skilled in economic forecasting and planning, in supervising personnel, and in organizing planning, production, and distribution systems. Performance budgets, stressing rates of production and their measurement, were developed to help managers apply operations know-how and research techniques in the assessment and comparison of work efficiency.

The need to improve budgeting became apparent as a consequence of many of the same circumstances which motivated the development of other system procedures. During World War II and thereafter men became deeply involved in extending their planning skills in order to develop technologically and economically the complex and powerful machines and instruments they found themselves capable of constructing. As they refined their planning capabilities and technology, they gained a new position of power over their environment, their human competitors, and their destiny. Within their newly found systems of procedures, they extended simultaneously their analytical skills and their synthesizing or designing skills.

In addition, men realized that, because of their new power to plan and accomplish, they needed to predict better the effects of their more critical acts before they destroyed themselves or their environment. And they soon became aware that they needed to conserve their natural and economic resources because the enormous costs of the complex systems men put in operation consumed vast amounts of money and natural or material wealth.

With experience and experimentation professional managers and their economic advisors and technicians further improved their perceptions and theories of budgeting and derived techniques of budget control. Economic laws such as the law of marginal utility were invented and/or extended. This law recognizes that economic choices require decisions of selection because resources are usually inadequate. But it realizes that these choices are relative and conditional, not absolute. It acknowledges the assumption that the order of needs and priorities in social system changes is influenced in some interdependent manner by human need satisfactions and satiation. And need-oriented actions or production systems are subsequently dependent upon the changing order of needs. It is now generally accepted that in an economic system output values or needs are relative, multiple, and balancing. Their rank order of importance tends to alternate and occasionally recycle.

A growing body of economic theories and rules of procedure derived from systematic experience directly influenced the process of budget planning. As technological capabilities and economic growth accelerated and as economic control became more critical, economists and accountants were required to develop a budgetary procedure which expedited the more rapid and precise convergence of the traditional input budgeting of funds and resources and the determination of output needs and plans of complex organizational programs. The gradual application of interdisciplinary system procedures to budgeting was the ultimate result.

In a manner paralleling the evolution of scientific method and the systematization of the natural sciences, the science of economics progressed from ex post facto explanation (reaction to random economic events) to limited experimentation and finally to extremely complex planning or budgeting and experimental model and theory building. Econometrics is basically a science of macroeconomic or large-scale economic planning and prediction.

A conservative logic underlies all types of economic planning, budgeting, allocation, and accounting, however. Economics is built on a concept of operationally finite or limited resources. The primary test or measure of an economic decision is to determine that the return for every expenditure of resources is greater than the cost. This implies a comparison of:

1. The increments or additional values to be derived from additional expenditures.
2. The enumeration of alternatives for which the available funds could be expended.

From these basic rules or laws of economical decision-making one can then infer that precise economic judgments entail a systematic comparison involving a single cost-effectiveness index or measurement criterion (a monetary index) associated with a prespecified common purpose or objective.

As economists, accountants and business administrators developed theories and deduced useful working applications from them, an administrative or managerial science emerged. A significant contribution of this science was planning programing budgeting; techniques of operations research and the development of management information systems were other developments of this administrative science. These processes all became feasible through the combination of economic theories, budgeting procedures, and general interdisciplinary scientific and systematic theories and technologies.

THE DESIGN OF PPBS

PPBS is a resource allocation system which incorporates a systematic decision network or system within its framework. Both planning and intermediate purposing decisions are coordinated with operational control decisions and input resource decisions in this organismic or synthesized budgeting process. Effective PPBS combines the theories and techniques of system analysis and systematic decision-making with budget planning.

Some program budget experts prefer to describe this type of comprehensive and integrated budgeting as PPBES, including the process of evaluation within the acronym identified with the program budgeting process. It has also been assigned the acronym PPBADERS by those who visualize systematic budgeting as including planning, programing, budgeting, analysis, decision-making (systematic optimization), evaluation, and recycling. However, for the purposes of this chapter, systematic budgeting will be refered to as PPBS or program budgeting, with the inference that all steps of systems analysis as indicated in the MARS model apply in its development. The simplest program budget relationship may be indicated as follows:

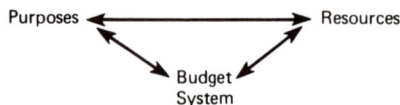

Purposes ←——————→ Resources

Budget
System

The Budget Planning Process

Human beings are unique in their capabilities for planning, for deciding a priori what is to be done and then converting the anticipated perception of the future into a state of present reality. Together with systematic experience (purposeful trial, analysis, and evaluation), planning is essential for advancing human inquiry and operational powers. As compared with traditional input and performance budgets, PPBS emphasizes planning or purposing as its most identifiable characteristic.

In traditional budget theory there are a number of ways for executing both the budget planning process and the process of controlling and justifying expenditures. Organizational policy makers or managers frequently constrain lower echelon management, budget, or accounting personnel in the planning process by establishing policies calling for open-ended, fixed-ceiling, increase-decrease, priority-listing, or item-by-item methods of budget presentation and review. Subordinates are given the greatest planning freedom in open-ended budgeting because they have no absolute constraints upon the cumulative monetary totals in this sort of budget planning. Fixed-ceiling budget restrictions set an absolute monetary total above which managerial planners cannot go. Increase-decrease budget processes give particular attention during budget planning processes to any item changes in a new budget which differ significantly from prior budget figures for that item. In increase-decrease budgeting the changes are given first priority in the reviewing and justification process. In priority listing, subordinate managers are asked to present all additional or new budget demands in the order of their preference or need. Item-by-item budget planning requires managerial justification of each input or output item requested and, if applied to budget control, it restricts expenditures for listed items to the amount specifically allocated. The above techniques of planning can be applied to the planning of any of the major types of budgets, input, performance, or programed budgets. However, in the traditional and performance types of budgets, they are applied piecemeal and independently. Only in program budgeting are they integrated into a comprehensive and synthesized budget planning system.

Planning programed budgets involves some other significant changes in the budgeting process. It shifts the emphasis in budgeting to strategic planning, to preparation. This means that the emphasis is shifted away from management or operational short-range preparation and from ex post facto input control budgeting. Thus the new budget emphasis facilitates a decreasing dependence upon operational managers, finance and accounting department personnel for making the original and major constraining budget decisions. The system of programed budgeting accommodates the increased value and potential of truly mission-oriented planning; it recognizes the increasing capabilities of human-technological systems. It changes budgeting from a reactive or ex post facto process or a pragmatic short-term market-oriented process to a more truly proactive or positive future-planning procedure.

Programed budgeting thus enriches the budget process by stressing an organismic or wholistic perception of an organization's major missions or programs. In this way it tends to liberalize or minimize traditional or present budget restrictions and constraints (natural conditions and human traditions) upon the organization and increase the belief in and motivation toward organizational growth and improvement.

Through the incorporation of system analysis procedures, programed budgeting emphasizes controlled or incremental change—not naive idealization, however. Systematic planners realize that extremely complex social planning requires that abstract ideals or global objectives be made explicit, operational, and verifiable before their attainment and value can be accounted for. Thus system analysis applied to PPBS adds to the possibility and probability that the budget can and will become a better controlling instrument as well as an integrated purposing instrument.

Yet programed budgeting is fundamentally macrosystematic in its orientation. It directs its planners to look first at the big picture, the long range. They can thus consider major or difficult objectives before they become immersed in the details of less important procedures or routine technologies, isolated operations, and minor costs. In PPBS the generic order of priorities are reflected in the major steps to follow in the planning procedure:

1. Define the organizational objectives (mission or program categories).
2. Relate these broad goals to specific programs (program subcategories).
3. Relate programs to resource requirements (program elements).
4. Relate resource inputs to budget dollars.

The reader will note that the PPBS process as stated is a simplified version of the planning and analysis phase of the MARS plan presented in the previous chapter.

Through a systematic accumulation of experiences in planning, managerial theorists have developed a number of rules of thumb applicable to all types of planning, including programed budget planning. These might be summarized as follows:

1. Identify the organization's major missions and work for persistence and motivation in achieving both mediating and long-range objectives.
2. Require that all goal statements are subjected to evaluative criteria which determine their possibility, feasibility, and economy.
3. Encourage participation by all informed decision makers, operational planners and budget managers as well as long-range policy planners, in the determination of both purposes and performance evaluative criteria.
4. Bring in fresh outside resources and ideas to stimulate and enrich budget planning and review.
5. In initial planning, encourage evaluative-free brainstorming. Initial errors will be eliminated systematically.
6. Record all suggestions, ideas, and results of planning research.
7. Allow time for reflection and evaluation after free divergent thinking.
8. Select an optimizing alternative only on the basis of systematic planning and decision making.

In the process of planning, a systematic program budget maker should be aware of past and present organizational and environmental conditions affecting his planning processes. He should be sensitive to the following planning variables and adjust his procedures to accommodate or ameliorate these constraining conditions:
Planning conditions may vary:

1. In the relations of the planner or planning unit to the operation being planned. The planning unit may be internal or external to the system being planned.
2. In the degree of fixity which restricts and limits the planning. Planning procedures may be quite deterministic or fixed or they may permit considerable flexibility.
3. In the degree of system and subsystem penetration or detail to be included in the plan.
4. In the significance of the operation being planned.
5. In the scope and geographic boundaries of the system being planned.
6. In the involvement of personnel within the system in the planning process.
7. In the prior history of general and related planning within the system.
8. In the time span of the planning project.

(derived from Dror in *PPBS*, pp. 102–106)

In the planning of all budgets, including programed budgets, the planning system is most often employed in an incremental manner, that is, the last year's budget or prior budget functions as the base model from which to work. Zero-based budgeting (no prior assumptions or models accepted) is a method of budget planning which requires initial planning and justification of all input and output assumptions and even methodological assumptions. It is therefore very difficult and expensive to operationalize and is usually regarded as a productive process only if it is employed very infrequently, perhaps every five years or so. As an important characteristic of PPBS planning, it is generally recommended that budget plans for more than one year are to be prepared, although these plans should be coordinated with regular annual input budgets which are also maintained.

Programed Budgeting

Since programed budgeting is a special application of system methodology, all of the characteristics of the scientific and system methods are reflected in this type of budget planning and programing process. Planning programed budgeting is a method or system which involves the following elements:

1. There is an orderly program structure. The organizational mission is perceived organismically. Its overall mission is the sum of all of its stipulated subprograms, which, in turn, are made up of component tasks and operations which are the direct output responses of input goods and service elements. The latter are convertible to economic indices.
2. Systems or programs are perceived to have both simultaneous and linear hierarchy. They have an alternating and balancing, independent and dependent, part-whole, and cause-effect nature. They vary in dynamic cycles from states of conservation or equilibrium (status quo) to states of growth and change (increments).
3. The system requires objectivity in the definition of budget terms, goal definitions, means definitions, and objective budget observation and evaluation processes. Programed budgeting is designed to systematically control subjective bias in budgeting.
4. The development of a programed budgeting system involves the theories and techniques of system analysis. This requires the evaluation of existing programs to see how well they meet present objectives and the determination and comparison of these pro-

grams and objectives with technologically workable and optimizing alternatives.

5. The budget analysis and evaluation system accommodates logical, scientific, political, historical, and normative criteria and means of verification and justification in addition to the usual economic indices. It facilitates prudent interdisciplinary decision making.

6. Quantitative aspects are treated quantitatively in programed budgeting.

7. The presence of qualitative aspects in budgeting is recognized as inevitable; judgments involving qualitative elements in budgeting are processed as systematically and sensitively as possible.

8. Risk and uncertainty are assumed components in all rational and systematic procedures and their productive applications. Analysis is recognized as essentially incomplete. Measurement of effectiveness is understood to be always approximate.

9. The budget instrument provides an improved basis upon which to develop and monitor an effective information feedback system, for controlling operations and sharpening operational decisions.

10. The budget process permits adherence to a time cycle. It meets current and emerging budget schedules and encourages long-range planning which optimizes both significant and consistent goal attainment and economic efficiency and permits prompt and systematic updating.

11. The total budget process produces both a descriptive and an accounting classification document which is meaningful and acceptable to all planning and operating personnel.

As many of the methods of system analysis suggested in the MARS model as is possible and practical should be employed in PPBS planning. Wherever possible cost-effectiveness calculations would be used to sharpen comparisons among new program alternatives. And alternatives to every major program and every program element in the operational plan should be offered in order to improve the probability of optimizing the system. In conjunction with systematic technological and economic analysis, planning procedures testing important changes or alternatives for sensitivity should be attempted also. And a social-political analysis of the major alternative plans is a component of careful programed budget planning.

When cost-effectiveness methods are applied in budget planning, economic managers and operations researchers realize that choices among output objectives or production goals in any system or plan are essentially a qualitative matter and involve human qualitative judgments. They understand the limits of their calculative techniques. They know that calculations assume a previously determined output objective and a common criterion or indexing system for determining the measurement process. The relative merits of alternative means-methods can only be calculated in terms of relative effectiveness in achieving an agreed upon objective.

As PPBS is deliberately macroanalytic in its intent, it acknowledges and in fact requires greater tolerance for uncertainty than is sometimes common within the more traditional accounting-type budgeting processes. Thus it recognizes the need to accept other means for decision-making in addition to operations research calculations and statistical analysis. Sensitivity analysis is an accepted procedure in planning long-range budgets. Some of the more speculative planning processes such as Delphic method, scenario writing, and other sequential guessing procedures are considered valuable for long-range or futuristic economic forecasting. Both extrapolative and interpolative processes of forecasting must be employed in program budget development.

Thus PPBS takes on the basic scientific and systematic characteristics of tentativeness as contrasted to the primitive human desire for absolutes. It penetrates boundaries and perceptions assumed to be fixed in traditional budgeting, where (1) the amount of money to spend is absolute no matter what goals and needs exist, and (2) where fixed objectives must be achieved no matter what the costs.

Budget experts have for years offered the truism that the budget should be considered as a planning tool and not an absolute and infallible end in itself, however. They have long advocated that budget systems or plans should not be presented or accepted as so perfect that they cannot be modified both in their subsequent planning strategy and in their tactical applications. Flexibility in the application of programed budgets as control instruments is particularly essential. This is true partly because the programed budget is based upon rather abstract or subjective programs, partly because the method is not as well tested in operation as other methods, and partly because lower echelon personnel need to adjust both their planning and accounting procedures to the system.

RESULTS OF PPBS APPLICATIONS

The evidence that budget experts are assembling from actual PPBS operations is limited and conflicting at this time. It appears that the PPBS process may be found most valuable in comparison with other budget procedures primarily because of the knowledge and skill gained in the improved planning and analysis operations it incorporates. Whether or not it will prove to be a better allocation and evaluation instrument or a more economical control instrument remains to be seen. It appears probable that traditional input budgets and managerial performance budgets will continue to be used in conjunction with programed budgets. Acceptable budget processes will not depart too far from observable or measureable inputs or performance outputs. Any type of functional budget process must facilitate the use of economic theory and operations research, statistical, and accounting techniques if it is to remain acceptable as an operational technique.

The overall effect of PPBS is to focus the budget procedure upon the determination of purposes and away from the restrictions placed upon accomplishment by management capabilities and/or by input resources. This could mean that the use of PPBS in large organizations might, because of practical logistics, tend to centralize organizational budget decision-making in a central policy group. Such centralization and bureaucratization might become economically unproductive if operational managers and technicians lost their freedom and desire to innovate, improve, and economize. Some experts feel that the reversal of the order of information flow in budget planning so isolates budget authority and responsibility that the probable loss of technical assistance and democratic support is a risk that could prove to be a significant limitation of the programed budget process.

It is generally recommended that present input and performance budget instruments be continued alongside any new programed budgeting procedures introduced. The input resource categories employed in traditional budget classification systems are rather well understood, and accepted. In addition, they are based on natural and/or rather determinate elements rather than poorly understood goal abstractions. They have an advantage of simplicity and concreteness, dealing with natural object categorizations which are easily observable, identifiable, and measureable.

Where PPBS budgets are developed it is recommended that general mission or program classification systems be developed first and that subclassification procedures or facet analysis go only as far as appears initially productive. Another suggested alternative for introducing program budgeting is to apply it to one selected program among all organizational operations before attempting it across the board. This latter procedure serves as a more penetrating accounting experience, but it does not provide much insight toward the development of budget categories which are simultaneously mutually exclusive and exhaustive of the universe set of organizational purposes.

Budget experts suggest also that, if a PPBS system is installed in an organization, care should be taken to adjust the new plan to the particular organizational system. Every effort should be made to keep the planning and controlling functions flexible, particularly if the plan is a long-range one. It is also recommended that considerable freedom should be allowed management and operational personnel in both budget planning and budget control.

Many expert accountants are less than enthusiastic about program budgeting procedures. They believe that PPBS values lie primarily in the planning process and suggest that these benefits might be obtained independent of the budgeting process. It is their belief that PPBS tends rather to augment and enrich traditional input budgeting than to replace it, because eventually PPBS categories must be converted into conventional cost categories in order to satisfy the needs of management and accounting personnel who desire and require input monetary and resource statistics.

Accountants perceive that many major limitations of budgeting procedures are not alleviated through PPBS. It is still difficult to quantify and secure output quantities which reflect operational goals and which can be quantitatively related to input classification and costing data. The allocation of proportionate costs in budgets often remains a gross estimate rather than an accurate accounting. For example, the allocation of general overhead and support services remains an inaccurate and subjective judgment process. The assigning and accounting for input and operating responsibilities in performance budgeting still requires gross assumptions of input-output or cause-effect relations.

Thus PPBS contributes little to existing evaluation procedures employed in traditional budgetary processes other than the definition of new criteria. Procedures for assessing validity and reliability are problems of measurement and research which are primarily outside the scope of the budget process per se.

Budget and accounting experts summarize the strengths and limitations of PPBS in the manner described below. If PPBS is used exclusively:

> Operating levels of management will not have an effective management control system.
>
> The accounting system will not produce cost reports which are useful to operating management.
>
> Cost accounting systems which already exist will continue to exist, resulting in duplication.

Other observations regarding the strengths and weaknesses of PPBS may be made:

If output budgeting proves not to be viable (and it may prove thus), it will have to be abandoned or overhauled at great expense.

It is probable that output elements will not satisfy the principles and standards of federal and state cost accounting structures. At least a conversion system linkage between the two types of budget classification systems is necessary.

Traditional budgeting tends to emphasize practical advantages. Program budgeting stresses purposes and goals. The two perspectives always must be reconciled in a truly systematic budget process.

Program budgeting seems to serve best as a managerial and accounting instrument when output products or services are naturally or physically categorical and deterministic (permitting easy either-or classification). It seems to work well in hospital accounting.

There is nothing magical about PPBS. It contributes nothing which cannot be obtained via other linkages between input budgeting and system analysis procedures. A total linkage between budget planning and comprehensive system analysis in complex social organizations is extremely difficult and expensive and is seldom attained.

At the present time the utility of PPBS for educational administrators remains somewhat uncertain. Since education is concerned with multiple and subjective programs, educational goals are difficult to define operationally and categorically. And educational missions or programs are extremely difficult to rank in order or to numerically weigh and combine. Whether PPBS will aid extensively in rationalizing educational budget planning and decision-making remains to be seen. In theory PPBS will become increasingly useful in education as education becomes a science and a technology. Almost certainly PPBS will be a useful by-product of system or organismic planning of complete and integrated organizational missions. Data processing technology and programing and operations research techniques and economic forecasting will probably accompany the effective use of other system analysis procedures in all of their useful applications to educational administration, including PPBS.

As the result of limited experience in the use of PPBS in educational organizations, some general knowledge of the side effects of the new budgeting system has been acquired. It has been found to be difficult to establish major organizational objectives which are simultaneously acceptable to all, operationally meaningful and measureable, and practical. Complex educational objectives are difficult to define, operationalize, analyze, compare, and evaluate. Goal displacement is likely to occur when trying to assess complex instructional programs. Complex cognitive objectives are sacrificed for more easily measured lower-level cognitive objectives. Affective objectives tend to be ignored or deemphasized. Observable teacher mediating output characteristics tend to be employed in place of actual program output measures in budget planning and program evaluation. The desired intent of planned change and improvement does not automatically result from the development of a programed budgeting procedure.

Programing educational operations is a complicated and subjective procedure. Properly designated program or mission objectives should be mutually independent of other stated objectives. An adequate enumeration of program objectives, in order to be capable of full rationalization and programing, should exhaust the universe set of educational program objectives. At the same time the list of objectives should contain only those programs which are mutually exclusive or independent of one another. And all subprograms should ideally be proper subsets of only one program or system higher in the hierarchy. Such a classification matrix or taxonomic ordering of output objectives becomes extremely complex. This quality of ordering frequently cannot be maintained productively when it reaches the level of decomposing goods-services inputs into monetary allocations, for numerous material goods-services resource inputs are employed in numerous subprogram operations. At this point the task of the budget planner is, of course, to subdivide the natural input resources into proportionate parts for proper allocation. This can be done, but the results of this degree of detail do not always appear to be of greater value than the costs involved.

Educational business managers who have become somewhat experienced in programed budget applications offer these observations regarding the utility of PPBS:

1. PPBS tends to require and improve high level decision-making through improved goal definition, the provision of options, systematic comparison, and the development of better networks for information feedback.
2. PPBS requires that educational program objectives be stated in behavioral response classification terms. This requirement will be opposed by persons resisting this educational perspective.
3. Instructional classifications employed in educational PPBS require that instructional missions and programs be decomposed into many subcategories which indicate classes of subjects, building, service units, grade levels, instructional units, types of students (their input state), the level of accomplishment to be reached, etc.
4. It is difficult to devise alternative operational systems for all service programs, tasks, etc. Therefore it is difficult to achieve edu-

cational benefit-cost efficiency through PPBS. PPBS is basically neutral in educational cost reduction. This is really the province of experimental and operations research and operational know-how.

5. PPBS is somewhat dependent upon the board of education and the superintendent for its success. It has other political implications also; therefore it must be considered and evaluated in respect to its social-political constraints and effects.

6. PPBS is best implemented where there exists an extensive and adequate data bank of planning, costing, and evaluation information. Electronic data processing and managerial information systems are useful instruments for developing effective PPBS. PPBS is no better than any other plan based on poor information.

7. There is a tendency for PPBS to improve planning and allocating procedures more than it improves accounting and evaluative procedures. It contributes little to the latter.

8. PPBS does not necessarily alter the organization and involvement of personnel employed in budget construction other than reducing the accounting domination of initial purposing somewhat.

9. It is frequently desirable to try PPBS on limited programs rather than across the board. Regular input budgets should be maintained in all areas converted to PPBS.

10. A system of reclassification which coordinates input and program budget categories generally is necessary for accounting purposes.

11. Effective PPBS development almost always requires the employment or retraining of personnel in order to obtain sound system analysis capabilities.

12. As do other system procedures, PPBS tends to increase the quantity of budget data before it improves the quality of the data. It is not likely to have immediate production or cost reduction benefits. Its initial benefits may be in the development of new attitudes and skills in the personnel involved.

EXAMPLES OF PROGRAM BUDGETING PROCEDURES

It is beyond the scope of this book to provide a thorough practical guide for the development of a programed educational budget. However, it seems appropriate to conclude this chapter with two illustrations of representative budgetary procedures.

In any educational budget development program it is necessary to construct a budget-making schedule or calendar in order to insure meeting the legal and institutional deadlines. Adequate time allotments and careful scheduling of budget-planning procedures are even more important when attempting a new type of budget development, especially if the new budget increases its time dimensions to include more than a fiscal year. A representative sequence of stages or steps followed in the development of a comprehensive budgetary procedure includes the following twelve steps:

1. A prebudget conference is held to outline the annual budget calendar, fix dates and responsibilities. Only the central administrative staff is usually involved in this step.
2. The superintendent and business officer or system analyst prepare the criteria for budgetary planning and the budget forms and instructions.
3. The business officer distributes estimate forms and instructions to all personnel to be included in the budget-making process. He prepares estimates of fixed charges and other nondepartmental expenses. He will also, at this time, prepare preliminary estimates of tax and other revenues.
4. Budget conferences are generally held at three levels: intra-departmental, intra-school, and at the central administrative level. Teachers, supervisors, department heads, administrators, and other personnel should be involved at the level of their expertise or interest.
5. Completed budget estimate forms are forwarded for incorporation and integration with other inputs at the appropriate level, gradually working their way up through the system. They are eventually returned to the business officer for consolidation.
6. Final budget conferences are held, involving the same personnel as in Step Four. Where problems appear, a cooperative solution is sought. In some cases justification data of expenditures may be required.
7. The business officer confers with the chief executive and/or the budget committee and they determine the amounts or programs to be finally recommended to the board of education. The superintendent also determines the elements to be considered as the official revenue estimates of the school district. The superintendent or his delegated representative prepares the budget message. The business officer usually prepares the final budget accounting document for submission to the board of education.
8. The combined budget document, including both a descriptive budget message and an accounting classification system, is turned over to the board of education for its consideration.
9. After appropriate consideration and changes, the board adopts or approves the budget.

10. The business officer or superintendent checks to assure that the final approved budget document meets all institutional and legal requirements.
11. When the budget is confirmed by all requisite criteria and personnel, the business officer posts the budget allotments in the accounting allotment file.
12. A post-budget conference is held by the superintendent and his staff to analyze and evaluate the entire budget procedure and make suggestions for future changes. This step could also be considered the prebudget conference for the coming year.

The problem of developing a comprehensive set of meaningful and assessible set of educational program objectives and subprograms is perhaps the most difficult task in the design of an effective educational programed budget. Initial efforts toward developing educational PPBS objectives have involved the use of standard educational taxonomies such as the Ten Imperative Needs of Youth and the Bloom and Krathwohl taxonomies of cognitive and affective objectives. These approaches toward the development of program budget categories have not been singularly successful.

An educational program budget classification schema which appears to have more practical utility is presented below. This schema is intended to be a representative rather than an ideal model of the broad categories or program missions which could be included in a program budget. The reader will note that the basic programs fairly representative of a typical elementary and secondary school district are stated in output terms rather than as input elements or mediating activities. Other than this, the classification system appears somewhat similar to a traditional educational budget schema. The particular budget classification system presented was derived from a model developed in a 1969 publication of the AASA entitled *Administrative Technology and the School Executive* (pp. 97–99). It has been slightly modified.

Budget Program Categories	Program Estimates for Fiscal Years					Total 1971–75
	1971	1972	1973	1974	1975	
I. Educational Growth and Development						
A. Cognitive-Intellectual Growth 1. Primary Service Centers						

Budget Program Categories	Program Estimates for Fiscal Years					Total 1971–75
	1971	1972	1973	1974	1975	
2. Elementary Service Centers						
3. Middle School Service Center						
4. Senior High Service Center						
5. Exceptional Student Service Center						
6. Adult Education Service Center						
B. Social-Personal Development 1. 2. 3. 4. 5. 6.						
C. Physical-Psychomotor Development 1. 2. 3. 4. 5. 6.						
D. Vocational-Production Development 1. 2. 3. 4. 5. 6.						
Total for all programs under I.						
II. Compensatory Experiences						
A. Dropout Prevention 1. Among Elementary and Middle School Students						

Budget Program Categories	Program Estimates for Fiscal Years					Total 1971–75
	1971	1972	1973	1974	1975	
2. Among High School Students						
B. Reducing Learning Deficiencies						
Total for all programs under II.						
III. Systems Maintenance, Evaluation and Design						
A. Staff Rapport and Negotiations						
B. Transportation						
C. System Security						
D. Evaluation and Design						
E. Fiscal and Material Resource Management						
Total for all programs under III.						
IV. Environmental Communications						
A. Parent and Community Relations						
B. State and Federal Relations						
C. Accrediting Agencies						
D. Attacks on Schools						
Total for all programs under IV						
Grand total for Programs I through IV						

If one were to use this type of budget program matrix as representative of the major missions of an educational system, subsequent system analysis would require decomposition or facet analysis of general programs into grade or unit level subprograms, subject matter and instructional service subprograms, etc. These in turn would be decomposed into task and learning units and means-methods-media inputs for which alternative choices would be found and compared as to their workability and economy. Analysis in depth of this kind is a complex and costly operation, but at the present time there are several efforts under way to complete just this kind of program analysis and budget development. The values to be derived from such an undertaking will have to be justified in the future.

12

Operations Management and Research

MANAGEMENT INFORMATION SYSTEMS
AND ELECTRONIC DATA PROCESSING

Some type of well-designed management information system is required as an essential component in the systematic management and
operation of any complex social organization. It is essential in the effective and efficient application of system analysis, PPBS, and operations
research procedures. Management planning, analysis, and decision-
making are no better than the quality and quantity of information and
facts upon which they are built.

Prior to any careful consideration of management information systems, it is necessary to distinguish among the functions of information
processing to be called in subsequent pages EDP (electronic data processing), general communications, and MIS (management information
systems). The term data processing properly refers to the collection,
tabulation, and manipulation of routine information, statistics, records,
and reports which are usually business accounting oriented and which
meet the legal and formal requirements of the organization.

General communications in complex formal organizations includes
all of the formal and informal reporting and communicating that takes
place in the organization. It describes the communications network and
flow of information that actually takes place when data is moved from
one decision point in the organization to another.

A management information system (MIS) is a carefully designed
system linking only the appropriate general communications and EDP
elements. It is a planned arrangement for assuring that appropriate data

is communicated in the proper form to the correct decision points at the appropriate time so that it facilitates organizational and managerial planning and operational decision making. Optimum MIS systems provide data relevant to futuristic planning and programing as well as information reflecting present operational conditions, including abnormal situations or exceptions to desired operating standards. An MIS system includes all data storage and classification files and systems, data retrieval systems, scientific research and reporting instruments, and all other modes of formal organizational information transmission and processing.

An MIS network ideally would include an optimum mix of men and machine interfaces, an integrated "total system" where the optimum information and information processing resources were applied to the resolution of every organizational information operation. The critical MIS design problem is to organize the flow, manipulation, and application of data in order to simultaneously facilitate optimum operations and maintain an appropriate data distribution, application, memory, and security network. Whether data processing, communications, or management information systems are dependent upon sophisticated machines or involve only human and manual procedures, effective MIS networks are dependent upon careful planning and organization.

Informed administrative managers are quite aware that a truly "total system" or comprehensive real-time or fully current information system is an impossibility. They are also aware that fully mechanized management systems are myths which will never attain full possibility or reality. These ideals can be approached but never fully reached. System managers realize that machines and preprogramed EDP operations can only manage basically routine, repetitive, and standardized operations; the capacity for planning changes in dynamic systems and the formulation of models for new systems require the relatively constant use of human imagination, intuition, and judgment. Fully mechanized information systems cannot handle organizational exceptions a priori (in planning) or ex post facto (in management by exceptions decision-making) except as the alternatives or branches have been previously experienced and identified.

At the same time, the astute managerial decision-maker realizes that the quality of human decisions and managerial strategies is highly dependent upon the presence of (1) relatively complete information and (2) relatively significant information. Thus they understand and appreciate the value of carefully planning and modeling management information systems. Productive planning has been found frequently to justify the relatively great expense it may involve. MIS systems have often provided a great human pay-off, particularly when the decision-making they enhance has involved a choice that is of an extremely critical, complex, or costly nature.

As a result of an accumulation of experiences in MIS, EDP, and other system procedures, administrators have become aware of the need for considering administrative philosophies and theories as well as techniques in the planning of organizational operations and related information networks. At one time bureaucratic theories led to the planning of information networks which centralized all important decisions in accordance with the "one big brain" concept of organizational leadership and decision-making. It was predicted that middle managers, including school building principals, would soon be unnecessary and obsolete components of organizations. Experience has proved these oversimplified bureaucratic theory applications to be wrong.

Modern interdisciplinary organizational theorists believe that organizational optimization is best approached by designing MIS systems which encourage decision-making to be decentralized to the lowest point in an organization where the necessary skills and competence and sufficient legitimate authority intersect to permit the most useful application of new information in the making of a meaningful decision.

As managers gain more experience with MIS and EDP they frequently decide that the human use of human beings in man-machine systems and interfaces is an economical and productive as well as an ideal principle to follow. In centralized decision systems the system carefully restricts authority in determining social and organizational goals and value considerations, emphasizing organizational hierarchy and status. Decisions at lower levels are routine and insensitve to human, social and psychological dynamics. In highly centralized and tightly structured organizations, human functionaries at lower levels are treated as production elements or machines, and they tend to produce at normative or minimal rates rather than at optimum rates. Their active human interests lie off the job.

In decentralized organizations hierarchy and status are deemphasized. All personnel have an opportunity and are in fact encouraged to participate in determining social and organizational goals and purposes as well as play an active part in operational or control decision-making. It is a principle of participative or decentralized leadership philosophy and theory that production optimization can only be achieved in the long run by maximizing participation in decision-making. The theory is that, as information is found to maximize and optimize production, organizations will tend to decentralize their MIS systems rather than centralize them. This will increase mutual goal purposing as well as facilitate constant, consistent, and current operational control. System procedures essentially distribute, standardize, and normalize both goals and means. It is only through decentralization that complex MIS operations can be fully integrated and simplified, making them balanced, effective, efficient, and economical for the long run as well as for the immediate future.

Thus effective MIS design and management requires and reflects the integration of many disciplines and considerations as do other system processes. Planning of complex MIS systems is a cooperative process and involves high level administrators, operational management staff and both MIS and EDP specialists.

Experienced system analysts suggest that many organizations would be better off to introduce organizational studies, system and procedure, and work simplification analyses before experimenting with EDP and MIS systems. They believe that a well-engineered MIS system is dependent upon a well-planned organizational decision network with functional and accurate job and position descriptions for its effective development. Computer and other data processing hardware and software feasibility studies should be developed only after operational structures are properly determined. They should not be allowed to determine the functions of human managers and operators in organizations. It is unfortunate that organizational analysis and planning have seldom preceded or accompanied MIS planning procedures.

Experience with MIS and EDP development indicates that the direction and management of information systems must be placed in the hands of a person with top management capabilities, one who understands management personnel and management problems in the particular organization and who has the ability to manage both operating and technical staff people. He must be able to apply administrative leadership and decision making theories of management, know the importance and methods of planning and evaluating, and direct the meeting of operational deadlines in an effective and efficient manner.

Good MIS and EDP managers have been found to be risk takers in addition to being highly organized and informed specialists. They must be flexible and capable of accommodating new plans and problems. MIS managers must constantly guard against internal and external operational and information biases. They have to keep currently informed on the organizational environment and the internal climate as well as on current MIS developments. As professional managers they must maintain an interest in organizational growth and improvement and in their own self-development.

It has been found that MIS and EDP systems seldom achieve a high level of production efficiency if they are designed for routine business processes and control decision-making only and never achieve any sophistication in managerial planning, research, and evaluation. Thus system procedures for planning and analysis, operations research and simulation, and basic and applied research capabilities must be tied into MIS systems and EDP if they are to expedite the optimum use of a multi-purpose system. Efficient MIS systems cannot be designed for fire-fighting or management by exceptions only. For this reason some organizations carefully divide their MIS operations into two departments or sections—a planning, research, and development (system) section and an operations section.

Effective MIS and EDP applications reflect the usual system benefits. Information tends to increase in quantity almost immediately. Its processing and distribution is more rapid and error-free. Data and facts are more likely to be presented and considered in quantified form. Adequate preparation for MIS and EDP tends to clarify what data exactly is needed and who needs it. MIS networks tend to increase the range of alternatives available to a manager and keep him currently informed regarding all operational conditions. Systematic EDP networks also provide more ready feedback regarding the impact of decisions. System procedures measure operations more reliably and efficiently and compare them with goals and standards more continuously.

A highly mechanized EDP system in an MIS network is often justifiable only if it leads to productive planning of new programs, operations research and simulation, or prediction of the relative benefit-costs of alternative programs. The cost of MIS and EDP multi-purpose systems must be systematically measured and justified in an efficient organization.

In the designing of MIS systems, an analyst must determine both qualitative and quantitative dimensions of the information and report network. It is his task to plan so that the data flow to a particular decision-maker reflects the nature of his organizational responsibilities and the scope of his decision making authority. For example, the higher the executive in the organizational hierarchy, the broader but briefer his feedforward and feedback reports should be. Information detail should be transmitted up through a management organization only so far as it will be used by the receiver. The ultimate information system design goal is to provide each manager with just the quality and quantity of information that he needs.

It is therefore necessary to design an MIS system so that the flow of reports is selective, relevant, and timely. A particular report should demonstrate effective presentation techniques which dramatize and accent its most significant facts or problems. Its form and language should be that most readily assimilated by the user. Conventional EDP accounting-type print-outs are not in a form usable to many busy managers and must be modified in an MIS distribution. All MIS reports should be consistent with their initial or formal purpose as well as their likely actual use. It is desirable to design an MIS network so that there is an optimum distribution of duplicated data and so that there is a minimum duplication of stored or filed data.

A well-planned MIS network analysis involves the following series of procedures:

1. An initial study of the organization as a whole is carried out.
2. An analysis is made of organizational information procedures and requirements and the establishment of MIS program objectives.
3. Systems, EDP methods and equipment are selected and an MIS department is structured. The system selected must be justified via a cost-benefit comparison with alternative systems and justified as the one best for carrying out the desired functions.
4. The system is designed for the specific MIS and EDP applications. In this step the scope of the network should be clearly delineated. Operational requirements are specified. A development and implementation schedule is established. Requirements for facilities, hardware, software, new personnel, personnel retraining, data conversion, parallel operations, and phase-in and completion procedures and dates should be determined.
5. The system is installed and there is a trial of all prototype operations.
6. The system is fully field tested and evaluated.
7. The system is put on line and procedures are set up for its periodic review and redesign.

Some system analysts believe that the "total system" emphasis or the attempt to install a multi-purpose integrated system all at once has been a deterent to MIS development and has been leading many MIS and EDP designers in the wrong direction. They recognize that complete machine management is an impossibility even at the control level. Machines and preprogrammed MIS and EDP systems can effectively handle only routine and standardized information operations, and an attempt to install a comprehensive system often delays the step-by-step development and progress that is necessary in many organizational changes.

In the development of MIS systems, a number of critical decisions must be made regarding the ordering and scheduling of particular MIS and EDP operations. Usually system analysts recommend that business and financial data processing systems should be installed first. Financial information is chiefly internal and historical. It is used primarily in a control manner and its timing is relatively critical. Although it must be communicated to top management, it is processed primarily in a specialized or staff department. Machines, programers, and software are most adequate in business functions. Therefore the business or financial component of an MIS network seems to be the most easily justified and implemented.

Personnel data processing and information distribution is another MIS function that can be easily developed and installed. It is also concerned largely with internal control and historical data. It differs from finance and payroll operations in that its timing is frequently not critical. Generally it is under specialized departmental control also, although personnel data must sometimes be communicated with top and line management. It is quite easy to install and it meets with minimum organizational resistence.

Logistics data such as procurement, production, distribution, and inventory information is a type of data which is still basically internal and historical. This type of data is of greater interest and concern to line and operational management, however. Thus it must be transmitted more widely through a network which includes many line operating managers. Logistic MIS networks require extensive line management participation in their planning if they are to be designed and applied effectively.

Typically in organizational information processing there are two general types or classes of information processing operations. General purpose information processing and flow carries data throughout the system in regular and dependable channels, often following a predetermined and expected schedule. Information of this type is assumed to be of general interest or to possess universal utility. Duplicated general purpose information is a form of mass communication. There is little capability or desire to maintain the security of this type of information once it enters the MIS network.

The second type of MIS or EDP activity is much more common and is similar to most operations management and research activities. This type of information processing and flow involves the batch program or off-line operation. Batch operations use MIS and EDP capabilities for the solution of focal or limited problems and often maintain considerable security precautions regarding the storing and transmission of this type of data. Batch operations include the application of stored information and special program routines.

Some MIS and EDP systems are designed primarily as multi-purpose off-line systems for batch operations. Such systems and machines often require the articulation of scientific and business computer capabilities and program languages and the careful scheduling of long-run one-shot scientific operations with a relatively rigid and inflexibly scheduled business routine. Most multi-purpose MIS and EDP systems are off-line or in-line operations. In-line systems involve temporary storage of data for computation and manipulation. Few multi-purpose operations approach an on-line or real-time capability. Real-time information applications exist in the form of special purpose controlling routines and systems which have limited purpose. Examples of real-time systems include: the guidance of high-speed projectiles, automated machine manufacture, transportation and reservation scheduling and communication systems, and limited aspects of economic and production transactions.

A system analyst planning an MIS network should recognize the inherent strengths and weaknesses of information systems managed by and articulated through information specialists and technicians and those managed and used by line managers. Information system specialists are a principal source of new ideas. They are a group not constrained by the need to show an immediate profit for their capital and operating outlays. Therefore they frequently are most interested in and capable of assessing change opportunities. On the other hand, system oriented MIS managers and operators are frequently resented and resisted by operations people, causing a delay in accepting responsibility for information and its applications. They often lack practical know-how and internal political sensitivity. Their concern for and interest in change often leads to an undervaluing of immediate production concerns and costs.

Operating line managers influencing MIS management have the advantage of possessing broad and detailed knowledge of the organizational work environment being simulated or analyzed and reported. They can expedite a ready acceptance of MIS information and are generally competent in persuading and developing on-sight manpower to use and apply new information. The disadvantage of too much influence by line management on MIS and EDP lies primarily in management's limited knowledge of system capabilities. They tend to overemphasize status quo work patterns and procedures and generally resist major changes.

After consideration of the above strengths and weaknesses of organizational line and staff subsystem biases relative to MIS and EDP management and operation it should be clear that the overall coordinator of information networks must be a top-level manager. He must be able to coordinate line and staff departments and operations in an effective manner.

Authorities on information technology such as John Diebold have long predicted that technological improvements in systems will assist in overcoming many of the presently perceived limits of MIS and EDP networks. Writing in the *Harvard Business Review* of September-October, 1965, Diebold predicted that future information systems would be more versatile and would more nearly parallel the natural and desired (ideal) flow of information and decisions in an organization. He foresaw that MIS networks would closely approach general purpose real-time and on-line capabilities. Their flexibility would increase and their unit costs would decline. Diebold predicted that big cost reductions and system capabilities would be affected by the expanding of random-access capabilities and memory units, permitting the integration of data used in planning and research with management control data. He forecast that graphic storage and visual display capabilities and the improvement of computer languages would facilitate easier communication between men and machines. Diebold suggested that the training of managers in system theories and techniques would greatly expedite the real contribution of MIS networks. As has been the case with many highly rationalized perceptions of the utility of system applications to social processes, it appears that Diebold's predictions are rather overly optimistic. Men and organizations seldom act as rational predictions indicate they could act.

An analysis of practical experiences with MIS and EDP networks indicates that these system elements are far from being ultimate answers to information problems. Administrators have found that organizational changes often come so rapidly that information system developments do not keep pace with them. Frequently individual managers have been found to lack defined job patterns and cannot articulate their routine information needs. In changing organizations and social environments, planning and predicting responsibilities make up an increasing portion of a manager's work, and the time and effort needed to develop elaborate information programs for this type of problem resolution is prodigious. In addition, the social environment in which many organizations currently find themselves is also changing so rapidly that internal information processing is of little value to managers and administrators. They need environmental and market information rather than routine internal information. Under such conditions heuristic or exploratory management simulations and scientific EDP procedures become more important than routine EDP procedures. Active and continuous communication between management decision-makers and system personnel is essential in such dynamic organizations. Machine or routine operations cannot determine whether a carefully calculated decision is the right one if the constraining conditions change. Experts often supply the right answers to the wrong problems.

While most organizations have not reached the level of MIS and EDP application that business and government have, there are a number of rules of thumb and operational theories that have emerged from extensive experience with these systems. As these may be of interest to educational administrators, they are summarized below:

1. In the design and development of an MIS or EDP system, an

analysis of organizational purposes and structure should be undertaken first, followed by the determination of MIS purposes and subsequent design.

2. Effective MIS and EDP generally serve to:
   a. relieve organizational managers of many clerical task hours and free them for more productive and creative functions;
   b. reduce calculating and transcribing errors;
   c. speed clerical processes and information retrieval;
   d. increase overall work accomplishment by both managers and clerks;
   e. reduce the impact of clerical personnel turnover;
   f. meet demands for more reports to county, state, and national bodies;
   g. save on reproducing and transcribing functions;
   h. eventually reduce some operating costs.

3. All system operations and applications are designed to serve. They have functional utility only and are justified only in their ability to optimize and economize managerial programs. They are primarily mediating devices which assist men in increasing their accuracy and rate of production, and generally do not replace human beings in making initial planning decisions and subsequent qualitative evaluations.

4. It has been found desirable to have a top educational administrator attached to MIS and EDP functions and departments and trained to the degree necessary for effective administration; this is preferred to the importing of trained specialists who do not understand educational problems and system needs.

5. Criteria useful in guiding the analysis, selection, and purchase of man-machine information systems include the following:
   a. The criterion of production efficiency requires consideration of the time required for production, the most effective use of professional control capabilities.
   b. The criterion of production economy involves consideration of unit production costs, initial equipment costs, costs of installation, maintenance, operation, training, and supervision.
   c. The criterion of organizational morale involves consideration of present and future personal and group social factors and perceptions.

6. Following managerial system analysis, including an analysis of work flow and decision making as indicated in job descriptions, the definition of MIS purposes may be attempted. Decisions necessary at this time include a determination as to whether the machines and systems will be centralized to serve the entire district or decentralized to serve individual school units. Admin-

istrators will have to decide whether machines and systems will be purchased or leased or services purchased.

7. When converting MIS, operations research, or data processing functions from manual or accounting machine operations to automated functions it is best to start with one process at a time. Frequently EDP business functions are the easiest to convert first. Mechanization and automation can be initiated with only a key punch and sorter or an automatic accounting machine rather than a computer.

8. The orderly conversion of management and business functions to more highly mechanized or automated procedures involves these processes:

    a. Review carefully the system, processes, and personnel now in use. Have a clear understanding of present costs and personnel time requirements.

    b. Plan for adequate time for designing the new system procedures. Involve both managers and clerical staff in every phase of planning.

    c. Implement the changeover on a gradual basis. A systematic conversion schedule should be set up. Manual production of the desired new data outputs just prior to machine production stimulates interest in and respect for the machine processes.

    d. Follow up and evaluate each phase of the first-run machine processes.

    e. Involve the operating staff in evaluation and encourage their suggestions for improvement of operations.

    f. Establish and maintain schedules for the conversion process. When installation is complete, set up an annual calendar for the systematic review of all system functions and operations.

9. Initial results from MIS and EDP installations often require increased inputs of planning and operating time and additional material resources. They tend to initially increase the volume of output data. Quality control and improvement of communications and data processing capabilities are more difficult to achieve and slower to accomplish.

10. It is usually prudent to keep present manual or tabulation systems in operation until new machine systems are thoroughly debugged, tested, and audited. Provide for such parallel operations.

11. Mechanization and automation of information functions usually pinpoint problems or shortcomings in present systems. This often causes some unrest and apprehension among present staff and requires nonthreatening in-service development.

12. "Canned" packages of machines, and record and report systems

or other software usually do not prove fully adequate for a particular MIS or EDP installation. It is necessary to assure that such systems are fully tailored to a particular organization's present and anticipated needs.

13. A MIS or EDP system should not be allowed to determine educational policies or administrative behaviors but vice versa.

14. Care should be taken to insure that MIS and EDP outputs meet all legal and formal requirements for reporting and auditing.

15. An administrator should not expect full usage or maximum efficiency from personnel or machines in any new system installation. Almost all installations are accompanied by an underestimation of the extensive work and time required to fully operationalize the system or network.

16. A log or file for each MIS or EDP development should be kept. This should record their predefined purposes and the history of their development. The file should include all charts, programs, diagrams, explanatory material and financial information associated with the installation.

17. In an MIS or EDP system the computer console operations should be carefully routinized. Generally it is desirable to:

    a. Maintain daily, weekly, and monthly computer operation logs.
    b. Prohibit changes in computer programs (debugging) at the console.
    c. Require reasonable uniformity in computer scheduling, coding, and reporting.
    d. Require that programs and systems be accompanied by complete instructions for handling malfunctions, input inaccuracies, operator errors, and program mistakes.
    e. Require that a console operator follow a standard plan of action when an operation "blows up" or fails to work.
    f. Plan operations so that there is a reasonable capability for increasing traffic or load volume when necessary.

18. School executives probably make even more special purpose decisions than business executives, for the range of educational problems is typically greater and the number of optimum or satisfactory solutions is much less standardized and routinized. Thus it would appear that the system procedures of greatest value to educational administrators may involve primarily a general understanding of the basic logic of system theories and systematic decision-making. Knowledge of the significance and strategies of determining precise objectives and developing a reasonable range of feasible alternative plans of operation should precede the development of precise standardization procedures

or programs and the mathematical formulation of routine solutions.

EDUCATIONAL EDP APPLICATIONS

It is not practical to inventory all EDP applications which have been used by educational administrators. However, a representative list of routine operations is provided below for the reader who is relatively unfamiliar with educational EDP capabilities. In general there is a much greater history and use of machines and software in business administration and control management data processing than there is in planning, research, and simulation. The latter procedures are much more difficult, time consuming, and expensive to develop, and they are difficult to justify in organizations such as educational institutions where personnel and financial resources are marginal.

I. Administrative Services

1. Business Administration
   a. Budgeting
      Accounting (income and expenditures)
      Appropriation journals
      Encumbrance reports
      Balances for monthly financial statements
      Warehouse withdrawals
   b. Purchase Orders
      Write purchase orders
      Purchase order listing monthly for board
      Summary cards used for budget accounting
   c. Weekly recap of appropriation journal
      By location
      Department
      Encumbered amount, expenditures, and balance
   d. Accounts Payable
      Warrants
      Warrant listing sheets
   e. Aging Report
      Purchase order written but material not delivered
      Orders delivered awaiting invoice for payment
   f. Revenue Report
      Monthly
   g. Photocopy

Invoices
Bids
Legal documents
h. Capital Equipment Inventory of District Furniture and Equipment
i. Warehouse Inventory
Stock status reports (weekly and monthly)
On order listing
Withdrawal requests
Deliveries received
j. Stores Accounting
Warehouse catalog
Stores delivery
Back order listing
Transaction registers
k. Textbook Control Records
Circulation
State textbook inventory
Basic textbook adoptions book
l. Payroll
Warrants
Payroll listing sheet
Monthly register
W2 withholding tax forms
Retirement statistics
List of withholdings for insurance, credit union, teacher dues
OASDI report
m. Personnel Accumulative Records and Listings
n. District School Elections (bond, override, etc.)
o. School Census Data
p. Lists of Parents, Registered Voters, Mailing Lists, etc.
q. Employee Identification Cards
Name and address cards
Date of employment
Location
r. Cafeteria
Payroll (same records as business office)
Purchase orders
Accounts payable
Accounting
Warrants
Listing sheets
Warehouse withdrawals
Financial statements

    s. Transportation
       Ditto Cafeteria Functions
  2. General Administrative Services—Superintendent
    a. Master Personnel Directory
    b. Credential Listings
    c. Years of Service
    d. Salary Schedule and Step
    e. Photostats
    f. Lists of Community Relations and Curriculum Studies

II. Instructional Services

  1. Principal
    a. Labels for mailing
       Open house
       Orientation program
       Seniors
       School week
    b. Listings
       Grade point average and special awards
       School enrollment list
       Eligibility lists
       Students enrolled in various special programs
       Student addresses, parent names
    c. Computation of Eligibility Exponents
       Listings of players for specific sports
       Listings of players by schools
    d. Lockers
       Master deck of locker cards
       Assign lockers
       Locker lists
       Change locker combinations
       Print locker cards
    e. List of staff and student birthdays
  2. Attendance Officer
    a. Student Enrollment Book
    b. Home Call Cards
    c. Address Labels (including ninth grade and new students)
    d. Attendance Accounting
       Master attendance deck
       Monthly attendance cards
       Compute and run ADA reports monthly
       Special accumulation reports
    e. Enrollment Reports

Balance beginning of school year
Boy and girl count
Growth report
Modified physical education enrollments
File dividers
Readmittance Cards
Listing of students living in specific areas
Corrected student programs

3. Director of Student Activities
   a. Identification Cards and Transportation Lists
   b. Labels for Preschool Mailing
   c. Student Enrollments
   d. Student Programs
   e. Corrected Programs After Change of Program Instituted
   f. Senior Lists (of students, parents, addresses)
   g. Student Eligibility Lists for Clubs and Competitive Activities
   h. School Election Tabulations

4. Guidance
   a. Registration (total school, grade level, courses, sections)
   b. Permanent Record Sheet
      Accumulative sheet for grades
      Supplementary sheet for test results etc.
   c. Student Programs
   d. Class Counts and Balances
   e. Grade Reporting
      To parents and counselors
      Grade recording tapes
      Lists of failures, incompletes, eligibility, scholarship, etc.
   f. Grade distributions
   g. Summer School
      Registration
      Programing
      Grade reporting
      Attendance accounting
   h. Testing Program
      Scoring, punching, and interpreting
      Special tests for departments and individual teachers
      Item analyses and graphs
      Computing test scores for percentiles, etc.
      Comparison studies
      Preparation of grade and section placement cards
      Listings of test results
      Labels with test results
      Preparation of testing materials and distribution of same

    i. Research
      Grading distributions
      Determining local norms
      Item analysis of tests
      Class counts
      Study of groups
    j. Photostats
      Transcripts
      Changes of program
    k. Physical Fitness
      Classification averages
      Listings
      Student activities
      Health records
      Counselor evaluations
    l. Special Enrollment Studies
      Students enrolled in more than one subject in a department
      Student breakdowns by elementary districts, class size, etc.
    m. Grade Point Averages
      Boys and girls
      Number and order of periods
      Grade levels
      Class enrollment
    n. Principal's Honor Roll
    o. Non-eligible Lists
      Student activities
      Athletic department
    p. Counselor Cards
      Referral records, purpose, time spent etc.
      By school and counselor
    q. Change of Address and House Numbers
5. Nurse's office
    a. Student Enrollment Lists
    b. Change of Programs Records
    c. Photostat Health Records
    d. File Dividers
    e. Immunization Record Cards
6. Continuation school
    a. Programing
    b. Attendance
    c. Reporting
      Parents
      Employers
7. Adult education

    a. Attendance
    b. ADA by Area and Subject
    c. Mailing Lists and Labels
    d. Programing

8. Audio-visual
    a. Sort Cards by Shelf Number
    b. Punch Teacher Name and Request Data
    c. Inventory and Purchase Lists
    d. Teacher Instructional Assistance Records

9. Buildings and grounds
    a. Compute New Locker Combinations
    b. List New Locker Combinations
    c. Alphabetic List of Issued Lockers and Combinations

10. Teachers
    a. Listing of Teachers by Schools for Association Purposes
    b. Final Examination Assistance, Test Scoring, etc.
    c. Selection of Students for Special Programing
    d. Punching of Testing Keys
    e. Test Item Analysis
    f. Grade Reporting
    g. Attendance Reporting

11. Library
    a. Catalog Cards
    b. Circulation Cards
    c. Shelf lists
    d. Labels

12. Instruction
    a. Application of Computer Math and Science
    b. Vocational Business Training

13. Students
    a. Student Scheduling
    b. Student Directories
    c. Student Body Elections

14. Elementary schools
    a. Similar Administrative and Instructional Functions

OPERATIONS RESEARCH

The term research means to search again. Research is a careful, systematic review of that which exists and has previously been experienced or known. Systematic research is focal in nature, seeking to establish particular facts or relationships. Traditional ex post facto or Aristotelian methods of research primarily were and are devoted to the deliberate creation of symbolic models and knowledge systems from systematic observations of natural events.

Experimental research is a deliberate human effort to accelerate traditional research. In an exploratory and heuristic manner it creates new operating models from hypothesized symbolic models or theories. It encourages limited risk and trial leading to deliberate or accidental creation and discovery of either real or symbolic models through the exercise of the processes of physical or mental induction. In this way it accelerates nature and evolution, thereby increasing human knowledge and control over nature.

Pure research is research which seeks to test, for the sake of advancing knowledge, a symbolic model of some observable or unobservable object or force. It does this by hypothesizing via induction and deduction a cause-effect relationship convertible to material reality that is representative of the theoretical model under consideration. Then it subjects the symbolic and operating models to logical and experimental testing. The immediate or utilitarian value goals of pure research are unknown and unimportant. Knowledge produced in pure research is the important factor in its justification.

Applied research seeks to direct knowledge of cause and effect conditions toward improving the effectiveness and feasibility of a particular process of use to human beings. As compared to pure research, there is greater methodological and utilitarian focus in applied research. It involves controlled experimentation designed to (1) more effectively and/or (2) more economically produce a new or better alternate human technology or methodology.

Operations research is a particular refinement or extension of applied research. It is concerned with the improved methodology or technology of applied research, but only as a prerequisite to its commercial application. Operations research is primarily concerned with the methodology of commerce and management—economical and profitable mass production and mass distribution of prototype models, technologies, and methodologies. Operations research is sometimes called managerial economics.

As the name implies, operations research is the precise, systematic study and development of operations in some type of production system. In its functional applications it is the epitomization of practical system analysis and design. In its dysfunctional applications, it generally fails to be a precise predictor of behavior or economy in extremely complex and quasi-rational systems such as social systems or organizations.

Operations research is restricted in its fullest applications to systems whose functions can be totally prescribed and controlled, where all input and output variables are fully accounted for either by assumption, enumeration, or estimation. As many systems or problems are not too well understood, operations research tends to be more useful and definitely more precise for modeling only those subsystem operations where all input-output value assumptions hold and therefore the necessary and sufficient conditions are relatively certain. The application of operations research to dynamic macrosystems such as organizations assumes the dimensions, rules of procedure, and limitations previously explained in the MARS system of general analysis and design.

Applied research generally answers the "how" question of technology; operations research concentrates on the "how much" question of operational cost-effectiveness. Operations research techniques are useful in making more precise determinations where, if a system's output has limited utility (for example, in a market), the calculation of the amount of input is critical and important. It also facilitates calculative decision-making wherein, if the output limits of a system are unknown or uncertain, one can predict the amount of input which is the best risk for maximizing the cost-effectiveness or profit index of the operation.

In the same manner as system analysis, operations research extends scientific experimentation and systematic prototype design into the domain of economic-political market feasibility. It combines in some meaningful and functional way scientific measures of material technological effectiveness with systematic measures of economic and political cost-effectiveness into a single mathematical measure or indexing system of comparative values.

Operations research is definitely interdisciplinary or metatheoretical in character, as are all system procedures and theories. It involves a general system perspective which includes the macrosystematic processes of philosophical and historical reasoning in order to determine long-range or ultimate values and the microsystematic process of applied economics (including advertising), politics, sociology, and psychology in order to determine short-range experimental or market values. It, of course, follows the rules of procedure developed in the natural sciences and related technology which in turn extensively employ the methods of applied logic and mathematics.

As a systematic discipline operations research had its origin in World War II. Its interdisciplinary techniques were first applied in a major way in order to provide systematic assistance in the coordination of radar equipment at gun sites in England. Because of its effectiveness, it was rapidly transferred to America and applied to other complex and expensive military planning programs and operations.

Operations research as a system process is theoretically unrestricted in its applicability. Practically it is much more restricted. Precise operations research models can execute the basic operations research functions of mathematically calculating the optimum decision among alternative choices only when they simulate limited or determinate systems, where the assumptions made concerning input and output value parameters actually hold. The operations research discipline is required to provide a decision-maker with a mathematical equation which expresses a determinate measure of effectiveness as a function of the alternatives (controllable variables) and the uncontrollable variables, the system's constraints and the environmental restrictions, in the problem. This is stated algebraically in the equation $E = f(CU)$.

As systematic research methodology shifts from pure to applied to operations research it accepts a different number and type of restrictions in an incremental or cumulative manner. Operations research accepts a great variety of restrictions or assumptions in its model parameters in order to increase the value of the calculated differences among the primary experimental variables or alternate plans and decrease the value of errors determined by variations among the uncontrolled inputs. It assumes that the specified alternate treatments are inputs which can indeed be prescribed (at least in the form of estimates). Ideally, operations research produces equations containing mathematical calculations to be used singly or in combination as final proof of the optimum decision to be made among specified optimizing technological alternatives. It quantifies the relations of preference among alternatives.

Since it is restricted to quantitative methods, operations research technique is generally applied first to well-known and well-controlled subsystem components in complex systems and subsequently extended only as far as available information and practicality justify their use. With experience and reflective thinking operations research discipline has evolved its rules of procedure to permit practical applications of mathematical decision-making procedures to an increasingly greater portion of the operations in rather complex production systems. Eventually operations research discipline attempts to formulate equations representative of technologically and economically indexed input-output operations in almost any macrosystem.

Complete mathematical formulation of a system in operation is called mathematical simulation. Operations research is particularly designed to simulate operating production systems, using some type of economic or monetary code or index system.

The experience gained in the applied science of operations research confirms facts and opinions acquired in programed budgeting and other system procedures. There are many problems and programs in existence about which little is known, and therefore little can be done to predict accurately future behavior or establish future operational control. Conscientious and responsible system experts readily acknowledge the in-applicability and impracticality of precise mathematical measurement for the "proving of" (prediction or reproduction of) extremely complex dynamic systems. They do not discourage the effort to improve forecasting or predictive methods, but they caution against assuming that quantitative predictive models actually represent all qualitative input-output variables in an extremely complex and dynamic system.

Gue and Thomas state the functional limits of operations research in these terms (Gue and Thomas, p. 299–300):

> After reflection on these assumptions one realizes that there is a vast amount of work yet to be done before the formulation and solution of decision models closely approximate real life.

A careful look at the assumptions underlying operations research techniques appears to confirm the opinion just quoted. In operations research decision-making one makes assumptions:

1. The decision-maker must have one or more objectives. Ideally they are fully quantifiable (the how much question of input and output is known or under the general control of the program).
2. In order that such decision-making be highly productive, useful, and precise, the decision-maker must have generated through prior experimentation, research, and/or experience alternate courses of action which are desirable (can effectively and efficiently attain the desired output state).
3. The researcher must have a satisfactory strategy of method (rules or criteria) for choosing among alternatives.
4. The procedure for measuring or indexing the effectiveness and the cost of the alternatives is known and accepted. A single index, often a monetary index, is employed in the measuring process.
5. The major premise of operations research is the premise of optimization. If there is a feasible solution there is a basic or optimum feasible solution.

In operations research procedures one assumes an advanced order of control or certainty to exist in the planning and management of a system. Fully predetermined or sophisticated and precise operations research models assume:

1. Optimal control—a single effect can be generally accomplished.
2. Optimal sequence—a best means-method or ordered cause-effect is known and can be established.

3. Optimal reliability—the system is perfected to attain satisfactory operational (technological or methodological) reliability.

If any of the above stages of control which are assumed or understood fail, the entire system (and its mathematical representation) fails. Under the above conditions of certainty assumed in operations research, the only significantly independent or uncontrolled variables are those which may be manipulated at the will of the decision-maker. (Educators please note!) The generation of elaborate and sophisticated mathematical models or predictors of human behavior can be challenged in application by the exercise of any counterstrategy which reverses or changes the assumed input variables and/or the desired output values.

One of the assumptions of system analysis which applies particularly to operations research procedures is that quantitative means are used wherever quantitative measures are possible and are likely to be more precise and productive. Operations research statements almost always involve statistical computations. Yet operations research technicians recognize that there are frequently subjective or qualitative components assumed in the quantification process and that therefore the final result is only approximate. The precision of mathematical formulas is entirely dependent upon the assumption that they truly enumerate and represent all significant qualitative and quantitative variables functioning in the system which they model.

In order to give the educational administrator a descriptive overview of the customary practices and inherent capabilities of operations research discipline, it seems desirable to present here a summary of fundamental operational theories and principles of operations research. The basic problem or program strategy employed in most mathematical operations research modeling is the linear equation or a sophisticated extension of this type of equation. Some of the basic assumptions or rules of thumb of linear operations research procedures are the following:

1. The assumption of proportionality—an activity or event is a "black box." Its internal behavior is unknown and generally unobservable, but its output state is regulated by the input state and activity rate in the system. A mathematical determination of the input and activity rates is possible if the output state is held constant, and vice versa.
2. Nonnegativity—in an input-output transformation or black-box event any positive multiple of a relation is possible. Negative quantities of an activity are not possible, however. A negative activity cannot exist.
3. Additivity—the total amount of input and output specified for a

system as a whole equals the sum of all subsystem inputs and outputs. The outputs of one system are equal to the inputs of the receiving system(s). This is called the material equation. The reader is reminded that this assumption is scientifically and rationally acceptable but realistically unprovable. There is actually some inefficiency or entropy, some lack of information in a system. There are many instances where the optimization of a subsystem does not bear a linear relation to the optimization of the macrosystem. The term *suboptimization* is used in operations research for instances of this kind. The terms *synergy* and *serendipity* are employed to indicate unknown positive output conditions or benefits in a macrosystem which appear to be greater than or in addition to the sum of all subsystem inputs which can be accounted for.

4. The linear objective function—this assumption states that, among all input-output elements functioning in a system, one element, the output objective function or goal state is regarded as "precious" or more valuable than one or more presently defined input functions or values (which therefore may be expected to be changed or consumed in a transformational activity). The quantification of the precious output element(s) determines the activity's pay-off index. Activities that require or consume this output quality in abundance greater than they produce of this quality contribute negatively to the system's pay-off or operational value. Those operations that produce an excess of the precious quantity contribute positively.

The reader will note that these assumptions are consistent with general explanations of system theory and information theory presented earlier in this book.

In the world of business management and engineering, the types of decision problems which are typically confronted and resolved by operations research techniques are the following management operations. Note the careful and subtle differentiations which are reflected in the stipulated definitions of these input-output variables.

1. Inventory problems—inventory problems are concerned with the control and maintenance of quantities of physical goods. As stated this is an input problem only and is independent of output allocation.

2. Allocation problems—allocation is concerned with the relocation of input resources. The allocation of a single resource is a problem of assignment. When more than one resource is allocated to ac-

complish more than one goal (output), the problem becomes one
of distribution. Quantity control and optimum mix are allocation
problems.
3. Waiting line problems—this type of problem is concerned with
or constrained by facilities (permanent space resources) which
are employed to house operations which may fluctuate in some
way. Waiting line problems involve the balancing or optimization
of controllable variable costs in the system.
4. Scheduling problems—scheduling problems involve the timing of
arrivals or departures of input units (including uncontrollable
variables such as environmental or client inputs) which require
service. Queuing discipline involves a special kind of scheduling
problem, the manner in which customers form into ordered
queues. Among the queuing alternatives are; balking, setting a
priority order, jockeying, and reneging—leaving the queue.
5. Sequencing problems—these are problems related to the establishment of an optimum order for servicing input units.
6. Competition problems—problems which involve an interaction or
counterstrategy between two or more active input or output units
with conflicting objectives. Competition problems are very difficult to analyze or decompose. In operations research they are
usually reduced to another type of problem which is less complicated.
7. Replacement problems—problems involved in replacing input or
output units which deteriorate.
8. Search problems—in the operations research frame of reference,
search problems usually involve complex decision-making. They
balance the sum of two costs: (1) the cost of decision errors, and
(2) the cost of collecting and analyzing data. The cost of planning,
research, and evaluation should be included in calculating the
total benefit-cost index of any system. Planning and evaluation in
any social organization is a search problem in the operations
research discipline.

In the modeling or programing of continuing or dynamic operating
systems, the operations research techniques applied almost always involve the use of decomposition techniques, wherein the process is broken
down into a series of steps for independent treatment. This pattern of
logical and mathematical analysis then requires some methodology of
resynthesis. Decomposition procedures often involve establishment of
one type of procedural control and its mathematical formulation at a
time.
There are a number of productive theories and rules of thumb gen-

erated via experience with operations research which can be applied to systematic educational decision-making. Operations research experts reason that system outputs or objectives may be either qualitative or quantitative. Qualitative objectives are treated as assumptions in quantitative methods. Quantitative mathematical decision-making may occur under a variety of qualitatively different conditions, however. The characteristics of these different classes of decisions are reflected in their categorization as decisions made (or applied) under conditions of (1) certainty, (2) risk, or (3) uncertainty. (Gue and Thomas, pp. 276–284) A more complete definition of these categories follows:

1. Certainty—in decisions involving certainty, the system's assumptive parameters are certain or closed. The alternate technology or operation is the principal variant. There are no significant random or strategic controllable or uncontrollable variables.
2. Risk—if each alternate plan of action or operation leads to any one of several outcomes, the outcomes will occur with known or measurable probabilities. This class of action assumes that the several models evidence behavior of some uncontrolled type, but that it falls within the range of calculated probabilities. It assumes that the order of preference can be calculated within some acceptable range of uncertainty. Controlled input variation is assumed in this category.
3. Uncertainty—in a condition of uncertainty it is assumed that each controlled action is uncertain. It may lead to one of several outcomes, and each of these outcomes occurs with unknown probability. In this condition there is uncertainty concerning the probability that the controllable variables therefore will in fact determine a difference which is significant and reliable enough to make a decision concerning the relative merits of alternate plans.
4. Extreme uncertainty—when risk and uncertainty are combined it is possible to have a condition wherein the amount and significance of both statistically controlled and uncontrolled variation is uncertain or unknown.

These four levels of variation of certainty indicate a decreasing order of available information and control. Operations research calculations are quite precise when they can be applied in conditions of certainty; their precision decreases as they attempt to use statistical measures to account for increasing problems of mathematical control. Logic, mathematics, and statistical methods are employed to calculate in conditions of risk and uncertainty, but the compounding of uncertainties will be reflected in the precision of the models.

Operations research employs mathematical strategies in the logical formulation and mathematical calculation of decisions made under conditions of great uncertainty, including strategic uncertainty. The techniques of gaming assume great uncertainty which can be represented mathematically by models which are based on assumptions similar to those used in the system analysis procedure of testing for sensitivity. Techniques sometimes used include (ibid., pp. 284–286):

1. Setting a maximum criterion—in this strategy the analyst imagines the worst possible outcome for one or more courses of action when the adversary is nature (random).
2. Coefficient of optimism—here the analyst imagines the best possible outcome for one or more courses of action against nature.
3. Minimax regret criterion—this model imagines a condition in which the analyst estimates the difference between the values of a given course of action and the value obtained had the decision-maker known beforehand what state of nature would prevail. (He is actually estimating what additional information is worth).
4. The principle of insufficient reason—the decision-maker assumes here that, if he is completely ignorant of the states of nature, he should act as if all states are equally likely or probable.

The applied theory of mathematical gaming extends operations research techniques well beyond the complex levels of logical implication and mathematical calculation described above. For example, many types of gaming assume the presence of an opponent who is capable of making competing counterstrategic moves. The problem now becomes one of mixed strategies. In multi-dynamic systems such as competitive social systems, there are many counteractions by adversaries which increase uncertainty in the system. And to complicate the modeling of social behavior, human beings are capable of caprice as well as counterstrategy! The general decision strategy in social competition involves the modeling of complex patterns of risk and uncertainty wherein regrets are minimized (the analyst establishes regret criteria or loss limits) and gains are maximized (goal or output gains are established).

Some of the basic principles of non-mathematical gaming and simulation, the modeling of multi-dynamic systems, were discussed in another chapter of this section. It is beyond the scope of this chapter to discuss all of the methodology of mathematical game theory even superficially. The interested reader will find an abundance of literature on gaming and on mathematical decision-making if he wishes to pursue his interest further.

In summary, mathematical modeling used in operations research must be capable of fully representing or estimating all of the prime determiners and functions which exist in whatever system or program it models. Operations research aims to produce a complete or closed mathematical equation which represents or simulates the system in operation. To be practical, the model must be capable of precise and rapid application if it serves a simulation control function. It must be practical and useful if it assumes a planning function. It is the opinion of many experts in the field of operations research and economic simulation that these disciplines are precise enough and practical enough to justify and perhaps almost require the extension of these planning procedures to more and more areas of important decision-making.

IV

ORGANIZATIONAL LEADERSHIP AND EDUCATIONAL ADMINISTRATION

13

Social Organizations as Systems

A social group or organization may be defined as two or more persons who plan or organize in order to seek and achieve specific goals common to its members. As operating systems, organizations therefore are composed of two or more extremely dynamic, self-regulating elements (people) interacting in events or series of events. An organization is called an informal group if it aims primarily at achieving a common but limited event goal or set of goals. When its aims become more complex and long-range, the group usually transforms itself by policies, legal codes, and specialized work assignments into a more formal or structured organization.

Some organizations may be maintained and sustained independent of original objectives. Such formal groups are called institutions. Institutions are formal organizations whose value systems have been displaced —diminished to become more realistic or amplified to become ideals and ideologies.

As a system, an organization or institution is a synthetic construct or element existing in men's minds only. Its linking structure or network is imagined, symbolic, and manufactured. Thus, only when the images of organizational goals or rules function to guide or control actual behavior of people can it be said that an organization or institution actually exists.

Since they are products of and in the minds of men, organizations are systems which are therefore entirely responsive to or reactive to and dependent upon their individual human components or subsystems. They are basically imitative of those normative individual human characteristics most significantly and/or widely distributed or communicated among subsystem personnel.

Organizations, as entities dependent upon individual cybernetic control, are extremely complex and multi-dynamic systems. Their state and momentum come from many input sources, and they are monitored or controlled by articulation and integration within the common communications code or network. Therefore, the flow of communications and actions among personnel and the exercise of control over the communications in an organization are generally dispersed among many personnel and can be initiated, altered, or blocked by any one person. Organizations are substantially stable or irreversible only by common consent and changeable or reversible by organized and persistent dissent.

Organizational communications and action involve individual closed-loop (cybernetic) behavior nested in a more loosely articulated closed-loop system. In all loops the direction of communications flow may be reversed; that is, the organization, through any of its subsystem personnel including clients and the environmental public, may either proact or react internally. The assertion that social change must come from some human source outside an organization is a myth.

Such a constant interaction or integration of dynamic elements would be catastrophic if the organizational system's interactions were not simplified or ordered by the establishing of discrete objectives and procedural policies which act as standardizers and buffers or regulators for and against the various levels and ranges of subsystem communications and actions. Organizational behavior becomes effective and efficient only by establishing, limiting, and specializing goals and means; by constraining the kinds and degrees of control or power of people over people; and by limiting the space-time-functional dimensions of the organization. But, in spite of all of these deliberate methods of simplifying an organization, they remain exceedingly complex probabilistic systems. Their dependency upon a human psychological base insures this uncertainty.

As organizations grow in size and scale, they tend to become more formal. Perhaps the most significant characteristic of formalization within organizations is the resultant normalization or displacement of original personalized goals and means. Organizations transform personalized goals into formalized, standardized, and depersonalized normative goals. They systematize and codify objectives, policy or planning procedures, role and job definitions, and rules and regulations controlling the behavior of subsystem organizational personnel. However, organizational rationality, like individual rationality, is dependent upon a teleological or goal-purposing system.

There is a division of labor in formal organizations—a division which can be characterized qualitatively according to scope (significance or breadth of function or geographic influence) and sequence (significance or persistence of influence in the time dimension). This qualitative dimension introduces and maintains the dimension or condition of hierarchy within the organization. Thus in organizations the personnel who make significant decisions organize themselves into a social-political power system of imaginary hierarchial social levels which culminate in a theoretically integrated and individually embodied unity of command or purpose which is centered with the chief executive and/or his board of directors. His position is the imagined focus of final integrated power and/or judgment.

TYPES OF ORGANIZATIONS

As organizations have been subjected to the systematic inquiry of political, sociological, and economic theorists, their qualitative and quantitative similarities and differences have become more clearly defined and interpreted. Today we realize that the differences among organizations include much more than size, location, and specified purpose.

In addition, we are becoming increasingly aware of the pervasive character of organizational existence in our complex society. Individuals soon learn that they often need to reinforce individual efforts and power with organizational capabilities in order to effect significant change in our modern social systems. Thus our highly organized modern society requires an individual human personality which is extremely flexible, permitting adaptive shuttling among the many group referents which interact with an individual. Our organized societies also tend to depersonalize the behavior of individuals in this way to the extent that a healthy personality in an organized society requires a high tolerance for frustration and an ability to defer intrinsic or personal gratifications from a considerable portion of his conscious social experience. In fact, not even self-initiated positive motivation and behavior directed toward socially recognized goals is a personal satisfaction easily experienced by individuals in our highly structured society.

When a society becomes so overorganized that it cannot accommodate human individuals and informal groups, it tends to foster more and more alienation and eventually social revolution. In order to regulate the interacting network of individual and organizational power centers, governments and governmental agencies are created to assist in the societal accommodation to progressive and productive change, wherever it is initiated.

As organizations tend to grow and to become more impersonal, rational, and efficient, personal satisfaction and achievement of members require constant adjustment to and acquisition of the normative patterns of social interaction and behavior. Such organization-based socialization is a significant part of modern living. Because of the pervasive effects of organizations upon individuals, educational leaders have a considerable responsibility to assist the young to understand this aspect of their lives. Personal choices of vocations and even avocational interest choices require an understanding of organizational involvements.

Etzioni (pp. 61–67) finds it convenient and useful to classify organizations and organizational behavior into three types—coercive, utilitarian, and normative. Each of these types is characterized by a general pattern of psychological interrelationships. Coercive organizations tend to produce alienation and/or total competition or rebellion; utilitarian organizations produce calculated or limited cooperation, bargaining, negotiations, and general indifference; and normative organizations generate cooperative communal idealizations, loyalty, and receptivity.

In coercive organizations such as prisons there is quite a different set of internal and external relationships of personal and group means and ends from those found in utilitarian business organizations, and from normative organizations such as schools, religious groups, the Communist Party, and, to a lesser extent, most governments. Organizational control and power in coercive or competition-dominated organizations is and deliberately remains unequally distributed. Social control over employees or operators in a prison is strict. Control over clients or operands (inmates) is unilateral and maintained by force. The scope of the control over inmates is nearly total. Sanctions in prisons are important; rewards are minimal. In most modern societies, highly coercive or totally competitive (monopolistic) organizations generally need some form of social license to exist.

In utilitarian organizations, goals, control, power, and reward systems are usually very carefully and deliberately calculated and negotiated. The game of social living in these organizations has rather well-defined rules and regulations. Utilitarian organizations are generally limited in scope and significance, and control of their decision rules is almost equally distributed between the organization and the client system. It is not equally distributed among employees, however.

The utilitarian organization is highly dependent upon its client or operand system; it shares all of the important decisions; it constantly initiates communications to its clients and reacts to the client responses. The effectiveness and efficiency of a truly utilitarian organization is dependent upon its ability to establish a market—a client demand for and commitment to its products or services. In order to accomplish this marketing task, it uses propaganda and persuasion on its clients or public.

Because of their dependence upon uncontrolled external variables (client needs and rewards), and because internal resources and rewards are distributed very unequally, utilitarian organizations generally provide little intrinsic satisfaction to most of their employees. They tend to separate the life space of an employee into work and nonwork categories, into either necessary or satisfactory categories of personal and social experiences. Partially for this reason, if utilitarian organizations attempt to extend the scope of their control too widely over either their employee operators or their client operands, the effect is generally negative. Laws, unions, and governments tend to function as instruments which limit the effect of utilitarian organizations.

Predominantly normative organizations are characterized by a consensus of purposes and values. They also stress a commonality of status and customary process as well as purpose. They generally produce a service rather than a material product, and they are usually quite concerned with the formal procedures to be followed by their employees in providing the service and even in the use of the product or service by the client. Normative organizations generally exercise a broader influence or greater scope over employees, their products, and their clients than utilitarian organizations.

In normative organizations the leader usually desires and is expected to exhibit his deference position—to be the official spokesman. He often employs his personal qualities, his charisma, and his expressive abilities to maintain his leadership position.

The displacement of goals (the valuing of charisma or of means or procedural goals over ends or product goals) is often so great in normative organizations that they sometimes maintain control over their clients or operands or their product following an initially utilitarian business transaction. Thus formal normative organizations, together with coercive organizations in particular, exhibit a marked reversal of the natural individual-social dependency relation. They tend to impose more restrictions on their employees and upon their client operands or services than are necessary or useful.

The type of personnel, market availability of employees, the system of rewards, and the personnel selection procedures tend to vary somewhat in different types of organizations. However, all organizations find that personnel selection should involve a correlation of personal characteristics with organizational characteristics. Proper personnel selection pays dividends in immediate and long-range production, in maintaining organizational morale, and in reducing the cost of subsequent organizational internal control. Post-employment socializing is generally considered a worthwhile organizational activity also.

In developing an understanding of organizational characteristics and typologies, theorists particularly stress the importance of the client relationship and the internal distribution of decision-making power among operators in the organization. The relationship between organizational employees and the client system or public that is served involves a production-consumption relationship—a market position for the organization. (Etzioni, pp. 76–91) Utilitarian organizations produce products and services which are quite dependent upon the external market, upon competitive cost-benefit factors. Their power to survive is strongly countervailed by forces outside their immediate and continual control. They are sensitive to (they react readily and promptly to) the client prior to a market transaction, but too much sensitivity to clients following a business transaction may be a handicap to a utilitarian organization. Normative organizations are only indirectly sensitive to client demands. Coercive organizations are deliberately insensitive to clients.

Other criteria may be employed to classify and differentiate organizations: Organizations may be classified according to size. They may be public or private. Primary mission (e.g., economic, political, religious, educational, etc.) describes an organization. Criteria of membership differentiate voluntary, contract, and conscript classes of organizations.

Blau and Scott (in Carver and Sergiovanni, et al., pp. 14–16) classify organizations according to prime beneficiary. They identify four categories of human functionaries associated with organizations:

Rank and file members or employees;
Owners and managers;
Clients (public in direct contact);
Environmental public (public at large).

Their system of classification of organizations is generated by determining which of the above groups is the prime beneficiary of organizational efforts. Where the employees or members are the prime beneficiary (i.e., the union), the organization can be called a mutual-benefit association. When owners and managers benefit most, the organization is a business. Service organizations are those wherein the client is the primary beneficiary, and commonweal organizations are designed to benefit the public at large. The basic procedural criterion associated with the mutual benefit organization is democratization. Business concerns seek efficiency. Service organizations stress professional service. Commonweal organizations insure justice.

The internal and external distribution of the types and degrees of knowledge and ability also affect the character of an organization substantially. In organizations with many highly educated or skilled professional members, internal power and control patterns are somewhat unique. Strictly professional organizations have as their primary product knowledge offered in some form of skilled service. Professionals, via codes of ethics, employ their knowledge primarily in the benefit of their clients. Thus they provide a service and commonweal function primarily, and only secondarily a system of self-reward and control. Therefore there is much that is normative in professional organizations. They are generally limited and specialized in scope, and they are relatively coercive in their employee-client relationships regarding their service function.

Because knowledge is a property which cannot be immediately bought or sold, arbitrarily reassigned or transferred within an organization, the professional or staff people in complex organizations are often protected, buffered, or separated from direct consumer-client contact or from the operating line managers in business organizations who cannot resist exercising their arbitrary political management habits. Staff operations in all types of organizations and at all levels are often separated from production processes and production responsibilities (line authority) because of this.

In order to simplify the problems which arise between professionals and line administrators in large or complex organizations, some non-professional utilitarian organizations circumvent the knowledge-control problems by buying most of their necessary specialized knowledge from the outside.

The reader by now probably has sensed that most real organizations do not fall neatly into the organizational categories advanced by Etzioni or by Blau and Scott. Rather they differ in the mix of behaviors characteristic of these classes. In semi-normative, semi-utilitarian, semi-coercive, and semi-professional organizations such as schools, a compromising utilitarian service function often supersedes the professional or normative function. Because schools have been primarily semi-professional rather than truly professional, because historically they have been administered by lay administrators, and because the effect of their service is long-range and rather subjective rather than immediate and objective, the power of knowledge possessed by practicing educational professionals has been less than perhaps might be expected. In addition, educational goal displacement and the subsequently inaccurate evaluation of personnel and production have occurred extensively. This has limited the influence of educational practitioners over their service and their clients. The influence of lay business leaders on schools and the resultant protective unionism has affected the schools as a service or commonweal institution also.

THE FORMALISTIC CONCEPTUALIZATION OF ORGANIZATIONS

Originating in societies distinguished by an unequal distribution of political authority and power, economic and physical resources, and knowledge, the traditional, bureaucratic, or formal conceptualization of organizations stresses the most determinate, depersonalized, and standardized of its characteristics—form, hierarchical power and structure, and the stability of the systematic elements and relationships (positions, status, rules and regulations, etc.).

In the most primitive of authoritarian social structures, the ruler and his position were indistinguishable. Authority was absolute and located in the hands of one or a few persons. The scope of control was great. Arbitrary control of both political and economic means and ends, actions and expressions, was possible. Control by fiat or decree was common.

Even today, formalistic or structuralist organizational theorists perceive position and status as dominant organizational properties. Authority and responsibility are highly centralized in the chief executive (unity of command concept) or in an executive elite. However, because of the extensive influence of democratic thinking and values, the present deference power and scope of authority of organizational leaders is more limited than formerly. Limits of legitimate authority, of position, power or control, are regulated internally through policies, rules, and regulations in all modern organizations; they are also regulated or restricted by governmental laws and informal social customs.

In most formalized organizations an elite called the governing board dominates the organization's legislative and judicial functioning. This elite determines all significant organizational policies. Administrative management, another elite, exercises the executive function. The chief executive is responsible for carrying out policy by organizing, specializing, and delegating it—by converting it to specified production goals and by developing rules and regulations which control all supervising and production operations. The administrative management staff also oversees any procedural adaptations necessary for adjustment to particular events or circumstances.

Within the formalistic organizational construct there is a rather sharp distinction between the official or managerial elite and the workers. The elite is usually expected to exhibit a more or less open commitment to the organization (to one another and to the hierarchy) and is usually rewarded on a deference basis which includes status, influence in the organization, or some other form of subjective personal reward in addition to more objective rewards. The myth that executives are really paid on an objective merit basis is often pure propaganda.

Executive elites are usually paid annual salaries (some have even longer contractual agreements) and occasionally are rewarded by means of highly selective and restricted merit or bonus systems. The working class or non-elite are paid for immediate unit production, sometimes either by the piece or by the hour. They are paid for observed and measured production only, as their commitment and deference value or status is perceived as minimal. The size and type of personal reward in formal organizations is related in a relatively subjective way to referent power (power of position, association, and persuasion), performance or expert power (utility), and openness of commitment or deferent power (normative conformity) of the individual employees in the organization.

Middle or lower management classes and professions, which traditionally have included school teachers and even school administrators, historically have been expected to exhibit openness of commitment together with prescribed performance and relatively low status without any adjustment of either their commensurate social status or their economic reward system. This condition indicates the traditional powerlessness of their position in society, as Eric Hoffer has indicated (see bibliography). A number of factors to be discussed later indicate considerable recent change in the social status and power of teachers and administrators.

Social or political power in tightly structured organizations is highly formalized or systematized. Official or authorized power and informal or personal power combine to determine the functional power of the official leader. In addition to the power delegated to the leader through his position or status, the leader exercises and extends his power through his activities. He shapes policies and controls the resources of the organization. These are his major command or policy decision-making functions. He also controls the public information outputs, the distribution of information within the organization, and the reward system. These are his primary operational or control decision-making functions.

In the traditional organizational perspective, a negotiated commitment is considered to be a highly restricted or centralized right. In almost all formal societies, even democracies, it remains a highly useful and responsible act. Such formal commitments are perceived as binding, so that the projected exchange of goods or services will be monitored or controlled by negative sanctions or punishments if they are not completed satisfactorily. Sanctions are acts of punishment administered to modify nonconforming acts toward future conformity. The general mores of organizations determine those boundaries or limits of behavior which are subject to the most severe sanctions. When formal organizations become less coercive and more democratic, sanctions are replaced by rewards as the major instruments of extrinsic motivation.

Within the formalistic organizational construct, the behavior of subordinate employees is perceived to be generally neutral in quality and passive in quantity. The task of the leadership elite is to manipulate this behavior in a favorable direction by the exercise of the system of rewards and sanctions. The best operational or control leader is that person who can manipulate the rewards and sanctions in order to secure the greatest benefits to the organization for the least cost. In such a frame of reference the employee without deference status is perceived as an object to be manipulated or replaced if a more efficient element or operation can be found to maximize the benefit-cost ration of the organization. In many, if not most, modern organizations this prerogative is still in the hands of management, but to a lesser degree. There are now significant counter-vailing mutual benefit and commonweal forces such as government, the unions, and social pressure groups, which compete with and challenge the social power of traditional hierarchical organizational leadership. Still, some employers have almost complete authority (including the prerogative of exercising personal whim or decree) to employ, promote, transfer, or remove lower personnel.

The formalist or bureaucratic interpretation of organizational design and function is premised upon the inference that conflict and competition in an organization are inevitable because of the economic theory of scarce resources and are in fact necessary and therefore desirable be-cause of their perceived negative valuing of human nature. It asserts that competition is a more realistic, more productive, and therefore a more important social force than is cooperation. The formal or bureaucratic manager believes that authority, influence, power, values, information, ability, and the desire to participate are all scarce resources, more un-equally distributed than equally distributed among human beings and that they must remain so. This traditional leadership attitude has been called Theory X by McGregor. It is also known as defensive leadership (Gibbs) and as bureaucratic or machine leadership.

The leadership role in highly formalized or closed bureaucracies essentially involves the activation or exercise of social power. The volume of initiating acts is the primary criterion of leadership in closed organi-zations. In return for this leadership effort (sometimes called the work-success ethic), the organizational leader is rewarded by relevant privi-leges and remuneration.

The inevitable alienation of employees in such organizations is be-lieved to be necessary by both the traditional bureaucrats and the modern structuralist group of administrative theorists. They believe that the bal-ancing forces in human-environmental systems and the standardized use of reward and punishment systems must occasionally, if not constantly, produce tensions through the competitive struggle for scarce resources.

The modern structuralists are fully aware that the degree of aliena-tion is of considerable importance, however. Yet they do not often recog-nize that the character and focus of the forces of interorganizational tension (i.e., whether they are truly goal-oriented or merely goal-displaced exercises of social coercion and control) are of equal or perhaps greater importance than the amount of this tension.

The function of knowledge and information in particular organizations is a cue to the latent perception of their leadership, authority and power structure. Highly structured bureaucratic organizations employ propaganda, advertising, and salesmanship (valuative-incitive expression based upon opinion, not consensus) to influence both their clients and their employees. Propaganda and persuasion as a predominant mode of communication in social units is more effective with the powerless, the uninformed or ignorant, and with those persons who already have a predisposition toward the vague value and demand content that comprise the particular propaganda (see Lasswell and Kaplan). Propaganda employed upon persons who are informed or are relatively independent of the existing authority or value structure is systematically effective only if supported by more substantial evidence. It must be at least partially amenable to logical inference and observational fact if it is to attain its desired effect for long.

In highly formal or tightly structured organizations or bureaucracies, official leaders, even if ineffective or inefficient, often sustain themselves in their positions by the use of propaganda and by the control of information networks. Some system technologists and technocrats suggest that computerized information systems will facilitate the firm control of information networks by increasing the potential and practicality of centralized decision-making and control. This seems to be an extremely short-range or naive perception of the actual dynamic function of information and communication in complex societies. It seems prudent for organizational theorists to question the inevitability of any social or technological arrangement or force which totally closes an organization. As Dennis Gabor, professor at the Imperial College of Science and Technology in London, has pointed out, ". . . exponential curves grow to infinity only in mathematics. In the physical world they either turn round and saturate, or they break down catastrophically." History seems to indicate that there is indeed a balancing and reciprocal interaction of the forces of individuals and social organizations.

CHANGING INTERPRETATIONS OF ORGANIZATIONAL LEADERSHIP AND POWER

There is little doubt that the development of a discipline of social psychology had a considerable effect upon the conceptualization of organizational leadership. The importance of informal behavior and communication and the significance of immediate and generalized social charisma (the art of leadership) and personality were defined and amplified by the first efforts of psychologists to study interpersonal behavior in organizations and groups.

As psychologists began to analyze organizational behavior, the categorical, stable, or determinate nature of influence became immediately suspect. Perhaps this is partially due to the fact that the social milieu at this time was becoming more democratic. The well-known Hawthorne studies were among the first research efforts of behavioral scientists to emphasize the dynamic cybernetic-psychological nature of individual behavior in social organizations. They indicated that even in formal organizations men behave according to norms and values extrinsic to the formalized social system and intrinsic to their own personal or informal group characteristics. Workers established *normative* production rates rather than *capacity* rates independent of rewards or sanctions. They sought and attained rewards other than economic and extrinsic, and they exhibited strong group affective identities.

The human engineering or mechanical-manipulative approach to organizational leadership and structure was discounted considerably in early psychological studies of group behavior. The research efforts of Mayo, Roethlisberger, and Dickson, and later studies by Kurt Lewin and Ronald Lippitt have substantiated the indeterminate nature of social behavior. Organizational theorists now realize that there are many functional overlays which modify the simplistic expectations of stimulus-response social interaction. Some modern structural theorists seem to have forgotten the pervasive character of individual psychological differences exhibited in social research.

But the human relations school of organizational psychology and leadership did not dominate organizational and administrative theory for long. Overemphasis upon the personal, the individualistic, the unsystematic, the exceptional, the informal in human behavior was characteristic of early human relations theorists in most of their social applications. They made little headway against the deeply entrenched structured authority of social traditionalism and formalism. The Second World War, the subsequent cold war, and numerous accompanying technological changes have mitigated against them. Some degree of system in social entities of any scale has always been necessary and desirable, and the social conditions of the last two decades have required considerable social structure. In addition, the hypothetical-deductive system of scientific inquiry and its derivative semi-closed structuralist organizational theory were generally applied in interpreting the quality and reality of organizational processes. Structuralism replaced human relations as the primary conceptual model for interpreting social organizations during the 1950s.

The structuralist theory of organizations represents a fusion of a number of interdisciplinary theories of social behavior. The disciplines of economics, politics, logic, sociology, and behaviorist psychology probably have been more instrumental in its formulation than were the more open disciplines of social and cognitive field psychology, philosophy, and the humanities and arts. As a theory of organization, the semi-closed system of structuralism forgoes both the absolutist perception of Aristotelian formalism and bureaucratic theory and the open, atomistic, artistic, character of individualistic humanism. It perceives an organization as a fairly well-structured, definable and predictable system of operations. Scientific objectivity characterizes and determines the ends and the means of the organization that are subject to study. Determinate structure prevails even though it is hypothetical or modifiable determinism.

Within the structuralist construct, goal displacement is once again the perceived organizational norm; short-term, operational, or process profit goals tend to dominate the organization's perspective. Organizational efficiency and maintenance have again become more important than personal effectiveness or personal efficiency.

The structuralist organizational unit is perceived to dominate the interaction between the social unit and the personal subsystems. The structuralist leader (the organization man) is other-oriented rather than traditional or inner-oriented. He is socializing rather than personalizing. In the structured unit, decisions are primarily mediating, competitive, compromising, and bargaining. This type of interaction supplies a reason for further decision-making. The neo-bureaucratic structuralist organization thus deliberately generates a dependence which requires the perpetuation of itself. The organization becomes an institution based primarily upon originally extrinsic or exchange values. In effect, personal values are the displaced or extrinsic values of the structuralist organization, although they are recognized. Goal displacement generally prevails.

There are slight but important differences between the structuralist interpretation of organizations and the open or general system interpretation of social units. The general or professional frame of reference, sometimes called Theory Y or participatory leadership in contrast to Theory X, the rationale of structuralism, stresses the dynamic, positive, and cooperative goal-setting qualities of the personnel rather than the assumed dynamic nature of the social unit itself. Within the general system perspective, the only justifiable reason for organizational existence remains the attainment of the limited and specified goals assigned to it by the mutual interests of all employee-client subsystems. Theory Y administrators perceive that organizations are humanly conceived tools for human use; they are necessary but not sufficient for human-environmental welfare. The value of organizations immediately and in the long run is entirely dependent upon the induced inputs of creative and inventive human subsystems as they interact with organizational resources or other human subunits. The individual human being is the self-initiating and self-controlling force, not the organization. Man's own synthetic creations (either his organizations or his technology) need not dominate him any more absolutely than nature has dominated him. Maslow, in his book *Eupsychian Management,* asserts the ultimate superiority of the general system or Theory Y perception of social leadership. Pusic and Gibbs (Carver, Sergiovanni et al., pp. 277, 316) foresee the necessity of increasing participatory or professional leadership and membership if organizations are to become more open and adaptive rather than increasingly bureaucratic.

Other than the difference of opinion as to who plays the primary and secondary roles in organizational behavior, general system, information, and cybernetic theories of organizations are quite similar to structuralist theories. Organizational goals are perceived to be pluralistic and variable. They involve relatively consistent hierarchies of product goals, but also permit the reordering of goals and priorities. Both perspectives hold that decision-making should be as close to the human elements possessing the necessary information and those needing or applying it as is possible and practical.

Consistent with his organismic perspective, however, the system theorist believes that organizational effectiveness is demonstrated not only in immediate production utilities but also in the long-term growth characteristic or the accumulative benefits of its production. An effective and efficient organization must both conserve and optimize all systems inputs (material and human) in order to achieve persistently desirable and economical outputs.

System theorists are very much aware that overconcern for immediate organizational efficiency or economy may interfere with organizational effectiveness by reducing the internal focus on production and by failing to exploit external environmental opportunities. They realize also that overconcern for organizational effectiveness may reduce external as well as internal organizational efficiency by destroying significant natural and human resources within the system, including the good will of employees and clients.

As a dynamic synthesis of self-motivated and self-controlled employees and clients, the cybernetically guided professional or Theory Y organization functions to achieve multiple and incremental goals. The value of the organization depends wholly upon its collectively perceived event-based and system-based synergy, its immediate and long-range utilities. In no sense does the system theorist assign any value to a formal social organization that is greater than the sum of its utilities. It is socially immoral in the libertarian sense of democracy and intellectually immoral within the tenets of logic, scientific, and system theories and methodologies to create a dependency of individual humans upon their own synthetic and nonexistent abstractions (including both ideas, beliefs, and knowledge and social laws, organizations, and governments) which is greater than the benefits they return. Where such a dependency is found, you will invariably find also a marked distortion of truth, a deliberate illusion. In such an instance a leader has used or is using propaganda to confuse his subordinates, asserting that his opinions and values are really the consensual goals and/or facts in the situation.

ORGANIZATIONAL RESEARCH AND EVALUATION

The uses of organizational research and evaluation and the results of evaluative efforts are a cue to the prevailing operational theory and an assessment of its relevance. Advancing organizational theory interacts constantly with research deduced from it. Thus it has become acceptable practice to expect that all high-level administrative research should have a solid foundation or rationale base in administrative and organizational theory.

Historically, bureaucratic theorists and formalistic managers employed their power of position to evaluate subordinates or other organizational components in whatever way they perceived that the situation permitted. Essentially this meant that they frequently advocated the perpetuation of a restrictive and arbitrary use of research and evaluation. Traditionally the evaluation of all elements in an organization was perceived to be a responsibility and right restricted to management; the method of evaluation was the sole prerogative of the executive. Unfortunately, because power, information, and an opportunity to participate were unequally distributed in most traditional organizations, the evaluation of personnel in particular was generally coercive and unproductive (almost inevitably so when it was unilaterally controlled). It seldom provided a true, mutual, or constructively meaningful experience or result.

In such formalistic personnel evaluation, goal displacement prevailed to a considerable degree. Methodological procedure, short-term production or profit, and conformity to the ideal employee image held by the superior were the results of many arbitrary and closed organizational personnel audits. At best a normative rather than an optimum production behavior resulted. The interpersonal relationships generated by such formalistic evaluation procedures produced a feeling of zero-sum competition.

The development and dissemination of system theory have changed both the purposes and methods of organizational research and evaluation, particularly personnel evaluation. Systematic professional evaluations of personnel stress the self-evaluation of individual behavior by each employee and mutual evaluation of the effect of this behavior upon the organization. Systematic evaluation emphasizes that the primary purpose of evaluation in all organizations is to assess the organization's total cooperative or systematic performance—the interrelationship of all parts to the functioning whole. An atomistic approach to evaluating any elements in a system, including personnel, is only justified if it eventually leads to a reliable unitary evaluation of the whole. Any lesser pattern of evaluation in a social organization is irrationaly arbitrary, socially immoral, and operationally normalizing rather than optimizing.

The professional system administrator emphasizes the synergistic value of research and evaluation. He attempts to use any and all evaluation as a positive rather than a negative force, and he stresses the desirability of informal, continuous, and immediate personnel performance evaluation over formal periodic evaluation. Above all, he works to build a mutual interest and concern (ideology) for evaluating the organization as a dynamic and evolving whole.

Systematic research and audit of organizational and administrative efficiency have improved in quality as well as quantity. Primarily through the use of system-based inquiry processes, complex organizational behavior is being assessed ever more comprehensively and precisely. In order to demonstrate the interrelationships between the system perspective and modern organizational research some of the results of modern research investigations are summarized below. In general the results of theory-based leadership research tend to support a generally open, cooperative, or professional leadership style as being more productive in a healthy organization or society, especially in the long run. Close control and supervision tend to produce normatively, not optimally. Results of research also tend to confirm the inevitability of equifinal or variable human responses produced by single administrative inputs.

There is an abundance of research literature which substantiates the fact that organizations depart from rational processes in all phases of their inquiry and communication processes. Organizations are adaptively rational rather than omnisciently rational in the information functions of sensing, measuring and encoding, information processing, regulation and control, and maintenance.

Cyert and March (Cooper et al., pp. 289–293) cite four critical overlays which modify classical rationality in organizations.

There is only a quasi-resolution of conflicts and deciding preferences.
Humans are characterized generally by uncertainty avoidance. Both individuals and organizations tend to avoid uncertainty or change.
Organizations are characterized by problemistic search (simple or

controlled induction). This type of particularistic search is dictated by specific and immediate problems rather than careful calculations of long-range expected returns.

Organizations periodically regraduate or transform their cognitive control or rationalization system. They learn. From on-going experiences they modify their procedures and policies over time.

All adaptively rational decision systems resolve conflicts only by establishing quasi-arbitrary preferences or priorities among alternative goals and optimizing choices among limited alternatives because of scarce resources (including time). This may cause a deviation from or conflict with long-range goals and optimum system attainment. Organizations have many ways of avoiding uncertainty and complexity. These may or may not be built into the rationale of the system. Employees avoid the uncertainty of centralized and arbitrary control by negotiating or arranging compromises, by imposing production norms and standardized operating procedures, by insisting upon traditions, and by demanding working contracts which limit the variation and degree of their commitment.

Systematic organizational research has hypothesized and affirmed many other characteristics of functioning organizations. According to Bass (Cooper et al., pp. 97–114), simple organizations tend to have the highest morale. Bottom-up (distributed or shared) decision-making shows more organizational gain and satisfaction in responses to a dynamic external environment than centralized or top-down patterns.

Research by Charles Bonini (Cooper et al., pp. 276–288) indicates that a business firm is generally most productive in an external environment which is highly variable and unstable. He concludes that the linear or continuous reduction of uncertainty and/or tension in an organizational-environmental relationship (a closed or bureaucratic relationship) is not always desirable. It tends to reduce costs, but it may also reduce production and sales.

Operations researchers and mathematical decision-makers in business organizations have made rapid advances in simulating and evaluating such systems. Brown, Ansoff, and O'Mera are three of the authors in Bursk and Chapman's *New Decision-Making Tools for Managers* who illustrate the utility of logical and mathematical models or linear programs which are useful in forecasting and predicting the effects of systematic organizational strategies and decisions.

Ruth Davis (Cooper et al., pp. 464–478) suggests that the simulation and automation of organizational information systems is not without negative values or functional costs, however. Computerization and automation may introduce problems of information security. Errors may be hard to detect. Automation tends to make the organizational leader more remote from his operating system, reducing his self-reliance, self-assurance, and indirectly, the organization's belief, faith, and trust in him. Davis suggests that automated systems should be evaluated as to their information timeliness, accuracy, availability (allowable distribution), reliability, completeness, and accessibility (availability to the requested user). Machine processes themselves can improve only information accuracy, timeliness, and perhaps economy. System analysis and design by humans are major determinants in the timeliness, completeness, accessibility, and availability of the information as well as its validity. All automated information systems should have their data flow subjected to program, mechanical, electronic, and manual controls and checks.

Systematic research in learning and training has confirmed many system theories and concepts also. Harold Borko (see bibliography) has affirmed the system concept of equifinality. He reports that the general order within any learning or training sequence may vary or reverse itself. Learning alternately proceeds from induction to deduction or vice versa. This finding should not surprise the system theorist who realizes that discovery and invention frequently vary in their order of incidence.

Rome and Rome have developed and reported a systematic analogue or descriptive model of leadership in social organizations in the book edited by Borko. They propose that an organization orients itself into two basic information subsystems connected by a unique network of communications. These systems are the *governing* system and the *technological* or operating system. (Some authors divide the governing system into two subsystems—a policy-making or planning system and an administrative management system.) Decisions in the governing system are called planning or command decisions, and decisions in the operating system are called control or programing decisions.

The technological system consists of all machines, messages, and applications of human energy that apply directly to the continuous creation or mass production of the products or services characteristic of the organization. This system gets the product manufactured or the service delivered economically. It is the system of information transmission, product manufacture, and distribution.

The governing system includes machines, messages, and personnel of another order. It introduces and controls change momentum in the organization and reallocates the technological components. It represents the truly transforming forces in the system.

Time and energy flow in different ways in these two systems. In the technological system the flow is primarily linear and systematic. It is more or less directional and continuous. In the governing system the flow is more cyclical and heuristic in form and structure. Behavior in the

governing system intersects or intervenes in all phases of the technological system, connecting and monitoring all of its operations at least
periodically. It purposefully deviates from specified directions, searching
and exploring for productive modifications in the organizational goals
and/or processes, or giving varying emphasis to different goals.

The technological system performs relatively atomistic and mechanical actions assigned to various definitive skill classes within the boundaries set by the governing system. The major properties of personnel
assigned to this system include the number of operators, standard skill
profiles, work profiles assigned to operators (missions, functions, tasks)
and individual variations in profiles (permissible ranges of behavior and
decision-making). The governing system controls the technological system by means of monitoring the completion of definitive work units, a
coordination of subprocesses, a reallocation and reevaluation of skills
and resources used during an operation, and the projected needs for subsequent systematic operations.

Decisions in the operating system set well-defined parameters or
limits for internal organizational operators or lower employees. Attainment of predefined production goals is the criterion of control efficiency.
Goal displacement predominates, emphasizing a focus on means and
methods. Economy is the rule rather than the exception in this system.

Decisions in the governing system are more complex. The most significant decisions here are unsystematic, idealistic, consensual, and
future-oriented rather than systematic, unilateral, realistic, and immediate. Command decisions involve both teleological and managerial concerns. They include directing the search for new goals, deciding on
significant changes, planning and programming for the adoption of innovations, directing the installation of changes, and evaluating the results.
The analogue of Rome and Rome is descriptively illustrative of the
general process of system analysis and program design which was developed more systematically and fully in the MARS model presented
earlier in this book.

Computer simulations of human cognitive processes have contributed
to the growth of learning theory and research. They have reemphasized
the importance of observation and analysis of the learning processes by
directly recording the behavior of persons asked to think aloud and report
the linear progression of their operations. Through their detailing or
programing of the step-by-step order of learning operations they have
produced more complex and precise and yet more consistent and manageable learning modes.

Human capacities are more accurately understood and described
than formerly. We know that, whether or not its activities are directed
toward conscious goal purposing and accomplishment, the human brain

can be characterized as intensely selective of focal activities. It is always valuing, deciding, simplifying. We know also that its most unique and important systematic properties are its extreme sensitivity, its complexity, its rich interconnections, its extreme activity, and its capacity to ascertain meaning (to use intuition) from grossly incomplete and disorganized data resources.

Research and evaluation have reinforced the conclusions of behavioral scientists that the generation of models, although not complete representations of operating systems, is a productive way of analyzing systems. Models are incomplete analogues of operations which often have more communicative capability and are easier and more economical to create than are complete prototype reconstructions. Mathematical models converted into computer programs have been found to have these primary properties: they often show a point vividly and simply; they serve as an archive or repository of knowledge; and they serve to affirm cause-effect relations logically in anticipation of their later material confirmation.

Computers used in the simulation of humans and organizations are demonstrating the importance of the affective-perceptive domain as a significant element in the understanding of individuals and social systems. Variability in human behavior due to belief, value systems and problems of motivation enter into all programs simulating interpersonal relations, economic systems, and strategies of competition in business, war, and diplomacy.

Systematic organizational research has provided theorists with yet another important conclusion. A careful review of the research on organization behavior indicates that organizational research is and must be a pluralistic and interdisciplinary methodology, employing every type of research method including philosophical and theoretical speculation, historical method, and quasi-experimental and experimental methods, ideally selecting the one method most appropriate to the particular situational goals and context. Thus a metatheoretical, interdisciplinary system perspective is necessary to provide the diversity of conceptual and disciplinary backgrounds necessary for assuring a dynamic progression of social goals and scientific-systematic operations.

Common operational methods applicable to the study of entire organizations include intensive case or historical studies of single organizations, comparative studies, laboratory experiments, experimental (iconic or prototype) simulations, and computer (synthetic and symbolic) simulations.

A researcher must select from these and other research strategies the most appropriate method for producing the desired information yield while at the same time meeting predefined standards of comprehensiveness, consistency, effectiveness, and economy. For example, laboratory experiments are precisely controlled attempts to ascertain fundamental processes and concepts via abstracted and predefined hypotheses. They usually involve a series of discrete trials. Experimental simulations, on the other hand, are more open-ended and continuous operations (they are more openly inductive or heuristic). They can vary considerably in their degree of complexity and they greatly enlarge the domain of search.

Computer simulations are a class of research methods which involve the invention and operation of synthetic mathematical models prior to their materialization in prototype operations. They are deductive-analytic, fundamentally closed or logically complete models of the phenomena being simulated. All variables including the dependent variable or solution and the range of alternatives must be accounted for in the process of formulating the model. This type of simulation may or may not involve human participation once it goes into operation because all decision rules are prescribed and decisions are made by direct calculation only.

Computer simulations vary greatly in richness and complexity. However, they must always include all pertinent (necessary and sufficient) structures and processes to be dealt with in the investigation, either as assumptive parameters, operating rules, or as heuristic or algorithmic processes. Computers can model human components only when they *do* perform as they *are expected to* perform, with predefined perfect consistency, efficiency, and rationality. Thus they are valuable for simulating the upper or ideal limits of a rationalized organizational performance rather than its actual or realistic performance. In general, computer simulations eliminate too much noise, error, or complexity from an organizational operation. They cannot fully predict natural or technological error perfectly, and human error or deviation due to conflicting value strategies or arbitrary or random caprice is most difficult even to estimate. Information concerning future behavior in social systems is always very complex and uncertain. Yet computer simluations can often offset a consequent reduction in the realism of their model by their speed and accuracy in simulating and ascertaining many values of the variables that are contained in their program.

The field study is an example of a semi-historical or ex post facto research methodology that is open at the beginning and closed at the end. It is a useful technique for researching a relatively untested or unexplored phenomena. Computer simulation, on the other hand, is the most sophisticated and refined of closed deductive-analytic operational research processes. It is applicable primarily where most of the necessary and sufficient information relevant to a problem and its solution is available or can be assumed.

The next logical step following a computer simulation or a thorough system analysis and design planning process is to validate the plan or program by testing it as a prototype model in a limited situational context. Following this, further cross-validation of the resultant prototype model in other real-life situations is necessary. This order of progress confirms the fact that an event in reality involves the transformed totality of any system and is always more complex than its preconceptualization or preprograming. Almost invariably the operationalizing of a simulated program exposes "bugs" that must be eliminated even when the operating model is a totally reconstructive application—a machine. When varying transformations or heuristic trial and error methods are involved in a simulated program the complexities of the situation increase rapidly.

Further elaboration and summarization of systematic organizational research and evaluation procedures and findings are generally beyond the scope of this book. Perhaps the reader has gained some insight into the interdependent relationships of purpose, operational theory, method, and conclusions that exist in organizational research and evaluation. Hopefully he remains aware of the fact that system techniques can usefully and economically resolve many practical organizational problems with or without computer hardware or software or elaborate research investigations. The system perspective can be applied to the logic of intelligent decision-making wherever it is needed, and intelligent decision-making can be applied to the simplest of human and organizational problems.

14

Social-Political and Economic
Power and Value Factors
in Organizations

The social sciences have been defined by Lasswell and Kaplan (p. xii) as policy sciences, sciences concerned with establishing organizational values or goals and in influencing or controlling policy decisions through interpersonal relations. Such political influence is characteristic of all social units. The strength and power of political influence is an important factor of control in all social groups.

POWER AS AN INSTRUMENT FOR ORGANIZATIONAL CONTROL

In the natural sciences power is defined as the rate of using energy and doing work. Natural or physical power includes the power of mass and the power of movement or momentum.

An eclectic operational definition of social power interprets it as the ability to do, to act, to influence. In a social context power is primarily the capacity to influence (communicate with and through) other persons. The political power of influence therefore exists only among interdependent yet self-regulating social system elements—people. Thus political power is characterized by multiple causation or input control centers, multiple directions of momentum, and multiple perceptions of its utilities and long-range values. It is therefore highly complex, variable, and unpredictable, having most of its consistency based in the compatibility of relationships and perceptions of groups of people.

As a perceptive-symbolic entity rather than a material entity, social power is doubly abstract and complicated. It is certainly dependent to some extent upon individual human feelings, values, and actions.

The art or science of politics as practiced by social-political leaders and propaganda or advertising experts is the art of persuading other people to act in certain ways. In social organizations characterized by prestructured authority and purpose, immediate and usually specific responses to political power acts are desired and expected.

Yet persuasive communications, no matter how subtle or how positive, are exhortative and valuative. They are valuative-demand statements or opinions of the maker and may or may not represent organizational consensus. The potential of these valuative-demand statements is dependent upon the predistributed or systematized values (the predictive code or shared ideology, utopia, empathy, and credibility common to the sender and the receiver) or upon the probable or imagined unique benefits which may accrue from the proposed action—the undistributed code or message. Operations researchers generally perceive the generalized and systematized value-demands as forecasting determiners and the more utilitarian, strategic, and economic value-demands as predicting determiners.

In a mature social entity the multiple facets of social power are integrated into a political *ideology*—a combination of myth and fact presently functioning to preserve, transmit, and reconstruct the social structure or event pattern—and perhaps also a political *utopia*—an idealistic myth designed to replace the present ideology. In a stable or closed social organization the ideology is well distributed; in a dynamic yet integrated organization the utopia is also widely shared. They are a matter of consensus, of public (distributed) rather than personal opinion.

The overwhelming American social ideology and utopia is democracy —an evolving form of social structure and control which is guided by three criteria. A democracy is a libertarian juridical commonwealth. As a libertarian system, democracy interprets the social relations between man and his organizations as value-ordered primarily in favor of the individual. As a juridical entity, an organization, although guided by laws and policies, may have any policy challenged and changed. And as a commonwealth, the policies and laws of the social unit are administered impartially. In a democracy the leadership is obtained by recruitment of a leadership elite based upon values to which there is supposedly equal access. There is equal opportunity for personal persuasion and for personal and social mobility. Within democratic theory, power—especially the power of information—is assumed to be highly distributed or equalized.

In a democracy there can be no authority and therefore no responsibility without consent. The exercise of such nonauthorized power is non-legitimate; it is evidence of primitive, naked, or coercive physical power. All assigned responsibilities and commitments among individuals and organizations in a democracy are mutually determined or voluntary and limited. All social and ethical commitments must be assumed and/or modified in a situation of free choice, not coercion.

Although the democratic ideal is identified primarily with governmental normative organizations, it eventually shapes human perceptions of legitimate leadership behavior in those coercive and utilitarian organizations which exist within such democratic societies. Too great a deviation from this societal ideal will not prevail unchallenged for long.

Thus we have in the democratic ideology of social organizations the perception of authority as proceeding from the bottom up, from the individual to the group. In the formalistic perception of organizational structure, we have a perception of authority as proceeding from the top down. The mutual exclusion of these two theoretical organizational power designs on some basis of organizational type or class is impossible. Some degree of resolution of these theoretical constructs must occur in all types of organizations existing in democratic societies.

And such is the case. Democratic organizations, as semi-closed and nested loop systems, permit some initiation of and/or reaction to the power of leadership wherever it may be located. The right of leadership is relatively independent of the person or position involved. Within highly democratic organizations the leadership momentum and opportunity is widely distributed and the reversibility of the direction of communications flow is great. The participatory or professional organization is consistent with an evolving and democratic communications theory.

Yet within all system theories there is such a thing as formal authority and power. It is granted representatively to occupants of certain positions by law, tradition, and consensus. However, within all modern theories of social systems and organizations, authority is limited and specific; its control rests with individuals comprising the system as defined in its charter, codes, and contracts.

Capable occupants of formalized organizational leadership positions soon learn to operate according to the codified formulas or uncodified norms expected of them. They usually are careful not to exceed defined or traditional limits too far. Fortunately, as leadership formulas are generally universalistic and vague to a considerable degree, the formal leader, who also is expected to interpret the leadership formula, has considerable freedom of action. Incidentally, the employee who is expected to demonstrate leadership in any form must have this same freedom of action.

The dynamic dimensions of authority, power, and leadership are a significant part of all leadership theories. System theorists believe that social power and leadership must be exercised to be maintained or to grow. In the system model of an organization it is expected that the locus of leadership power in a particular social situation will vary according to both the prestige and the demonstrated or exercised expertise of all persons concerned.

It is interesting to note that within our present perception of society as a highly dynamic system, power-minded individuals or power groups are working constantly to acquire authority or legitimation (power of position and ability) or are seeking to exercise effective influence over legitimate authorities. In such a dynamic society, when authority and power of control (functioning power) have been in the same hands for some time, a weakening of one type of power tends to weaken the other.

History seems to indicate that the individual who values power per se over power as an instrument for higher purposes does not remain as a social leader for long. Modern organized societies limit the misuse of power. They generally react against nonauthoritative, nonrepresentative use of power—nonlegitimate acts or overtly coercive acts by persons in authority and power acts exercised by persons not in authority.

In any organized society a position of leadership or influence can be characterized by its value status or power potential. The primary base of social power in an event or act of influence is the value which is a condition for participation (Lasswell and Kaplan, p. 18). In an event designed to influence others the initiator or leader must formulate in his power or value demand statement (1) symbols identifying all present power positions and potential, (2) the power base or event goal, and (3) the expectations relevant to the benefits for the person and/or the environment of the respondent.

In any single act of leadership or influence, the systematic value distribution, the initial degree of empathy, credibility, and belief, significantly affects the outcome, as the proposed act may be supportive of the existing state of potential or it may conflict with or interfere with the status of persons involved. Values are conflicting, facilitative, or compatible as their acts of valuation are perceived originally and as they are ultimately determined.

The perceived status of a social power relation or act of influence or leadership in a particular situation can be shown in relation to its stable or systematic characteristics (predistributed code) and its transformational or event characteristics (its message). Figure 11 below illustrates the degree of mutual agreement (informational compatibility, credibility, and facilitation) concerning the perception of an act of influence and its event goal that might exist.

Thus a well-planned power act or act of social leadership or influence involves ascertaining present value bases of persons and the availability of organizational resources. It also requires predicting the value position and potential following an event, and it involves developing a rationale and a methodology for carrying out the operation successfully.

Event power relations
(concern for effect or gain)

		Power equivalence	Power difference
Systematic power relations (concern for maintenance or loss)	Power equivalence	Systematic inter-dependence. Compatible cost-benefit perceptions	Mutual agreement of present status or fear of loss. Incompatible in-formation about gains.
	Power difference	Mutual perception of probable gain. incompatibility concerning cost or loss of present status.	Systematic aliena-tion or indepen-dence. No mutual linkages.

Fig. 11. System-event relationship.

The *realism index* of a social operation is the probability or credibility of the goal expected in relation to all other preestablished value criteria (possibilities). The *actualization* index is the state of a value as it approaches its predetermined potential. (Lasswell and Kaplan, p. 59) The realism index is the predictive criterion measure of an act, and the actualization index is the evolving value criterion measure.

The *security or expectation index* or legitimation of a social act is the product of the reality and actualization indices. The *economy* of an act is the synergistic maximization or optimization of the security index, the ratio between the total benefit of an act and its total cost.

Just as there are multiple input and output values functioning in most social or organizational operations, there are multiple power bases. The two major classes or bases of power are the class of political, deference, or status values, and the class of economic, welfare, or utility values.

As a social organization builds an evolving security index, an expectation or credibility condition, through a series of successful events, compliance to the legitimate leader's will and to organizational policies will increase. The affective or motivational set of all personnel reinforces the available knowledge of possible and probable meanings in such a situation. Mutual interest in the organization increases. Positive attitudes involving sentiments of faith and loyalty develop. More of the whole personality and interest of participants is involved in such a system. The organization becomes internalized in the minds of its members. This condition of open involvement and commitment is considered the ideal morale state of interpersonal relations in a formal social unit. It is seldom attained or maintained in complex organizations where there is no ideological or utopian image or where continuous face-to-face contact and mutual decision-making is impossible. It is most likely to be attained in a simple organization such as the family.

But even where such effective use of social power is impossible, a concern for social influence must inevitably focus on its human consequences. Ignoring such consequences will, in time, affect organizational morale, commitment, and performance.

The degree of synergy, satisfaction, or social esprit in an organization is the degree to which individual and group goals are accomplished effectively and efficiently through positive motivation and subsequent positive perceptions of values attained. Once experienced and understood, organizational synergy can be systematically maintained only by maximizing the opportunity for involvement in organizational planning and decision making. Policies involving system-wide consequences must be widely discussed. The maintenance of employee security and credibility positions and group affections must be preserved. Morale cannot be separated from voluntary self-initiated and self-controlled group participation. The attainment of maximum synergy in any social organization is dependent upon its function as an integrated multi-personal cybernetic system. Undistributed sharing of decision-making in an organization, whether exercised indirectly through policies and rules or directly through formal leaderships acts of persuasion, will have normalizing rather than optimizing results, except where the gain in utilities is immediately recognized or where such gains become positively accelerating (take on an obvious growth characteristic).

In the opinion of some theorists a systematic perspective of organizational leadership is fundamental to appreciating, understanding, and effectively exercising social leadership methodologies. For this reason all educators have a responsibility to teach about organizational leadership and to provide direct leadership experiences. They also have the responsibility to act in ways representative of good leadership behavior— to demonstrate their beliefs.

As organizations, schools have experienced until very recently and perhaps still experience a relatively simple pattern of administrative leadership. Their scope of influence or power over clients or operands (students) was and is normative or average. They have little economic power. Their organizational membership generally includes only one type of professional person. The quality or value of their product cannot be easily ascertained. And they operate almost as production monopolies. The semi-professional nature of schools tends to separate instructional employees somewhat from great consumer control. The primary basis of organizational control in schools has until recently been a strong and usually quite well distributed and communicated simplistic ideology and utopia, linking the school administration, its employees, and its clients to one another.

Bruce Biddle (Cooper *et al.*, pp. 150–172) has made an interesting and perceptive analysis of the American public schools as normative social organizations. Among the conclusions Biddle reached is the belief that the schools no longer have either a generally perceived or a clearly defined purpose or set of purposes. They exhibit extreme goal displacement. An emerging concentration on immediate or intermediate goals tends to restrict their long-range purposes and the resultant educational evaluation. Leadership action and leadership evaluation subsequently concentrate on teacher (operator) behavior and immediate overt pupil behavior. Thus an emphasis upon obvious behavioral restrictions and standards becomes the dominant organizational control focus, whether it is meaningful and educational or not.

In such a short-range, arbitrary, and entropic (unsystematic) environment, any substantial degree of significant change or proposal of change in long-range purposes is resisted both internally and by the client system. The end result is something of a chaotic system breakdown whenever anyone attempts significantly to improve the organization. Consequently there is a tendency for all personnel in the educational system to employ the bureaucratic devices of formalized communications control and communications distortion in the form of vague and meaningless propaganda to protect their status. Insisting that important educational decisions can be made intelligently only by professional educators is an obvious example of such status preservation.

Thus the traditional school tends to become increasingly bureaucratic, to maintain the status quo. Professional public relations statements are nebulous and/or unfulfilled generalities. Leadership is devoted to internal maintenance and to resistance to change. Other side effects of this power-value state of confusion or entropy are evidenced in the poorly defined educational goals and the equally poorly defined positions of organizational employees, their vague job definitions, work assignments, and procedures for evaluation.

Although Biddle's analysis is perhaps exaggerated if interpreted to imply that these conditions apply only to schools, there is probably much truth in what he has written. The interested reader may wish to interpret his own school in the frame of reference advanced by Biddle.

For many reasons relationships among teachers, administrators, board of education members, and clients of educational organizations are changing. Historically exercised paternalistic and authoritative patterns of lay board control have been challenged by both employees and clients (the lay public as well as students) during the last few years. More highly trained professional teachers and administrators are seeking to exercise their recently acquired positions of influence (based upon expert or informational power) to control or influence educational clients, while striving to negotiate in groups for more control of policy decisions involving organizational goals and processes as well as rewards.

The traditional simplistic and institutionalized imagined ideology and utopia within the schools in America are no longer very effective as an organizational controls or operations parameters. There appears to be a general depersonalization of relations among all social elements in the school and an emphasis on immediate personal economic or utility concerns. The politically informed educator is using organization and propaganda to gain a new power position for himself and for his profession.

It is interesting to note (although it should not be considered as unexpected) that the teachers, who now perceive themselves the dominant power group of employees in the schools, are seeking to maximize all of their value and power benefits at a minimum in cost to and commitment by them. This reflects a utilitarian, socio-political, competitive bargaining attitude. Teachers are finally learning the political lessons which utilitarian lay leaders (business and political) and structural theorists have been so anxious to teach them.

There seems to be little doubt that a slowly evolving redistribution of economic and political power, together with the increasing power of knowledge, is changing even the institutionalized structure of American schools.

It is often said that form follows function. This is somewhat the case in organizational design. When the power or leadership base of an organization is democratized (distributed) there is an accompanying change in organizational structure. A flat organizational design, reducing the hierarchical levels of power, generally results. This simpler, more democratic structure functions particularly well where the organizational values or purposes are widely disseminated and where information concerning the probabilities of their attainment is also widely shared.

Involvement in planning tends to flatten an organization also. Group decision-making at the executive policy-making level as well as at the operational controlling level tends to reduce the hierarchical levels of power in the organizational structure.

In a dynamic society characterized by a high degree of organization, it is essential that some transformation of power and influence must take place almost continuously. If social transformations are to be evolutionary rather than revolutionary, the society itself must create some structure for planned obsolescence control and for the replacement of formal leaders. Mechanisms and instruments for converting or accommodating divergent individuals or groups must be established. In interpersonal organizational behavior, we call this process *cooptation* or the absorption of protest.

Protest absorption is particularly important and necessary in normative organizations, including schools, particularly when such organizations operate on marginal resources, as the scope of control is fairly wide
and the morale problem of particular significance. It is given less attention in coercive organizations such as prisons or in "wild" utilitarian
organizations.

Ruth Leeds (Cooper et al., pp. 115–136) states that organizations
tend to use three primary modes of treatment of nonconformists: con-
demnation or punishment, avoidance or expulsion, and protest absorption. Of the three, only protest absorption has a direct positive benefit to
the organization.

It should be noted in this description of protest absorption that nonconformist behavior is not surreptitious deviant behavior. A nonconformist does not hide his dissent. He overtly expresses his interests and
exhibits the motivation sustaining them. Nonconformists often demonstrate both personal aspiration and ambition and organizational concern.

The optimum social coalition, cooptation, or absorption of protest
begins with ascertaining points of agreement of values and interests
between the organization and the individual nonconformist. The organization then structures a reasonable and limited (controlled) opportunity for the individual to succeed or fail in his endeavor while still
under the general control of the organization. While this requires a
mutual commitment of resources it may have mutual benefits. An opportunity to fail (or to succeed) is often the best instrument for protest
absorption. It may release hostility and/or gain substantive benefits to
the organization.

An active policy of protest absorption or cooptation in an organization will tend to create organizational synergy by building in obsolescence
control. It will expedite organizational growth, change, and effectiveness
while maintaining or improving organizational efficiency. Educational
administrators should consider the possibilities of initiating positive
programs of protest absorption rather than waiting for and reacting to
formalized and legalized procedures for handling grievances.

David Mechanic (Cooper et al., pp. 136–149) has studied the behavior of lower participants in complex organizations. His observations
may be of considerable value to the educator who attempts to understand
and influence his employees in a positive manner. Mechanic notes that
the individual power of laborers increases in relation to their immediacy
of access to supervisors. It also increases in correlation with a factor
which he calls consistency probability—the freedom of action brought
about by a significant length of time in service within the organization.
Consistency probability generates a subtle power factor of trust, certainty, identity—all security or expectation index or ideological factors.

Mechanic also notes that the power of lower participants is actualized through the manipulation of dependency relationships. Lower partici-pants learn to obtain benefit-cost trade-offs of privileges for obedience and compliance. They learn to negotiate or deal in informal exchanges of deference or even utility values in addition to their formal commit-ments. Mechanic, like most other organizational theorists, concludes that organizations need not and probably should not employ naked power (overt coercion as perceived by other than the user) to maintain their effectiveness or efficiency. This noticeable display usually accentuates the awareness of a power imbalance. Such an awareness has a direct negative effect upon the morale of the entire organization as well as that of the individual being coerced. It has an indirect systematic effect on organizational production.

THE FUNCTION OF VALUES IN ORGANIZATIONS

A human value begins as the perception of an element or event or the symbol of an element or event. It becomes meaningful only as human belief, experience, and judgment establish an ordering or binding relation between that which is valued and the valuator. Meaningful human values involve affective-cognitive perceptions, belief and credibility fixations, in which man orders an object or event in terms of (1) its rela-tion to him and (2) its comparative utility in relation to or in exchange for something else. Thus the event of valuing or decision-making in-volves a personal awareness and comparison of two or more external objects or conditions and the making of a decision which orders or ranks alternatives according to one or more criteria. Human values are therefore either criterion-referenced, personal, and intrinsic, or norm-referenced, impersonal, and extrinsic.

As the degree of meaning concerning valued objects increases, human evaluators are able to distinguish more precisely the qualitative, criterion, or significant dimensions and the quantitative, relative, or consistent dimensions of the objects being compared and exchanged. Human verbal and mathematical symbols are generally employed to ascertain and communicate judgments of these more or less meaningful value objects and relationships.

The concept of human values and valuing has no meaning except where there is a desire or need to choose and an opportunity to choose among objects or events. Animals seem to have such a self-initiating ability to a slight degree. But only man has this ability to any great extent. He can order not only immediate choices, but long-range, remote, or abstract choices as well merely by exercising either his memory or his imagination. The computer, with its prodigious memory, can aid men in this respect, but it contributes only indirectly to man's imagination, and ability to idealize. Rather, it is dependent upon these particularly human capacities.

A peculiar characteristic of human valuing is the difference between intrinsic or criterion valuing (generating purposes) and normative or extrinsic valuing (generating economical means-methods). Man, as the subject of his own valuing system, has an identifiable bias in his own ordered relationship to all external elements. He does not readily or rationally exchange his own value status or potential or even his lesser means, action, or operational values for a less favorable condition if he does not foresee the probability and possibility of personal benefit. In social systems dominated by intelligent and wise humans, the intrinsic value of human individuals is generally recognized.

As ordering or systematizing characteristics of human behavior, values coordinate the process of decision-making and shape or structure the behavior of social groups or organizations. Organizations are in fact designed to reconstruct and distribute the commonly perceived and defined personal values of its subsystem personnel. As they become formalized, organizations exhibit relatively determinate structures of normalized, standardized, and systematized human values. Formal organizations are founded to maintain and extend these ideologies and utopias as well as to attain utilitarian material needs. They translate the consensus of values.

In organizations in which social power and values are naturally and/or deliberately controlled (imposed or withheld) independent of general social involvement, there is considerable value displacement and value conflict. In such circumstances important values may be ignored or an object or experience may be over-valued in relation to its immediate use and its long-range benefits. For example, the leadership or management dogma that what is in men's interests—whether they are interested in it or not—is all that need concern us morally (is of social value) violates democratic and intellectual values. (Mills, p. 194) This dogma is a hold-over of past eras of absolute or formalistic authority, coercion, and social manipulation. It is not consistent with our knowledge of the natural universe or of human affairs. Therefore it is inconsistent with cybernetic, professional, or participatory leadership theory and with a system philosophy of ethical organizational leadership.

Values have their immediate or event-based characteristics as well as their systematic or long-range characteristics. Both the particular conditions of an individual in a situation and the stable conditions of the situation determine the possible value ordering, the true conditions and probabilities of an event. The determinate, systematic, or universalistic conditions of both an individual (system) and an environment determine the controllable and uncontrollable variables of an event interacting with and transforming a system. The most universal of human values are those determined by (are dependent upon) persisting environmental factors including common (consistent) or predominant (significant or unique) social preferences, and by idealized socialization and operational procedures.

A valuing act by an individual has its effectiveness and efficiency dimensions, its benefit, criterion, or gain characteristics, and its exchange, norm, or cost characteristics. Actualizing processes or technologies involve mediating or input cost goals, values assigned priority over immediately present states and conditions. They in turn are of lower priority in the system hierarchy than the preprogramed purposes of the action and are therefore exchanged or expended in order to attain these higher order values.

In primitive civilizations both individual humans and social groups tend to emphasize immediate survival or welfare values. The environmental situation dominates the human valuing and decision-making. Utilitarian or intermediate welfare values almost completely determine the behavior of men in societies concerned primarily with survival.

As organized societies develop and become more efficient, value concerns become more complex. The privileged leaders, figures of power, control the value structure of their organized societies. Through the exercise of social control, they manage the distribution of material resources and even the communicated propaganda intended to affect the perception of natural or persisting human values. In order to maintain the control of their resources and their subjects, they develop systematic processes for social control through communications control and the exercise of reward and punishment procedures. As most of the resources and power in primitive societies are not widely distributed, political leaders readily retain control of information and understanding. Fear and threat based upon physical power are augmented by fear based upon superstition and ignorance.

To some degree the reward systems in all organized societies also are based upon both material necessities and ideology (psychology) or myth. The conditions of personal status, credibility, and image (all deferential values) usually function in combination with material and welfare needs to determine value perceptions. The complex value systems which emerge tend to maintain societies in a reasonably steady state of dynamic equilibrium in which there is a simultaneous or an alternating balance of conservation and change forces.

In complex societies, particularly in authoritarian societies, religious or political indoctrination was and is the primary method of communication, education, and social control of normative values. Personal ideologies and utopias of the leaders or shared ideologies are employed to control and order the society. The most stabilizing leadership characteristics or norms of societies are the consistencies of leadership behavior or their substitute which are acceptable to the masses—the personal charisma, the consistency of the communicated propaganda, and the laws fixing the ideology in the system.

Propaganda in the form of doctrine and dogma has been employed for centuries to maintain social continuity and control beyond the immediate presence or even beyond the lifetime of a legitimate leader or government. Since the dawn of civilization, organized social units have developed social value doctrines, laws and codes which transcend individual rulers and facilitate internal and external social communication. Although democratic or inquiry-oriented societies often permit some degree of indoctrination and use of propaganda, their ultimate determination of important social values is achieved by exercising an orderly system of social transformations, either a true consensus or a voting pattern, ideally based upon considerable information, experience, and understanding.

Man long ago developed a very useful instrument for facilitating and measuring social value transformations or exchanges—money. This instrument, usually called the medium of exchange, possesses little intrinsic value but generally can be characterized by a relatively consistent or universal exchange value index. The value of money is determined almost exclusively by humanly perceived consistency and by its normative, universalistic, impersonal scaling or measuring qualities.

The practice and the science of economics is the discipline which is concerned with instrumental or extrinsic exchange values. It is concerned with facilitating the exchange of objects and services among people. However, systems of economics and their medium of exchange, money, are generally better predicters or measures of material or welfare values in a society than of the deferent, psychological, or political values. Ideally money should translate or mediate among all values.

Economic theory and methodology are concerned with maintaining both our complex adequate human exchange value systems and our long-range socio-economic growth and change. They have departed from classical conceptualizations based upon fixed or absolute terms and conditions toward scientific-theoretical interpretations reflecting hypothetical and evolving characteristics. The classical economic model in which money has an absolutely fixed value and where it is to be used as an absolute index for comparison among all entities, is no longer maintained. For this reason among others, organizational leaders, recognizing the shifting and tentative value of money, must augment this reward-exchange system by deferent conditions, properties, and privileges other than money if they wish to maximize social and organizational effectiveness and efficiency.

Although the purposing behavior of individuals may vary and although the perceived exchange value or benefit-cost index of an object or event may vary also, behavioral scientists believe that conscious and purposeful human behavior always reflects or is equivalent to a value need, a profit, or economic motive. Man, as a cybernetic system, inevitably acts to extend his status and resources or to reduce the loss of his status or resources.

Profit is sometimes called an index of social workability by economic theorists. In any socio-economic transaction it is assumed that the two or more persons involved both anticipate a benefit from the arrangement or transaction although the benefit may not be equal or identical. Their initial and subsequent willingness to participate is an indication of their anticipation or evaluation of such a profit.

The relative size of a profit in proportion to cost and to share of gain is significant in facilitating value exchanges. Anticipation of a personally unprofitable value exchange is generally nonmotivating.

Although our democratic capitalistic society encourages the profit motive for utilitarian organizations as well as for individuals, it also protects our client-consumers. Any extreme differentiation between cost and exchange prices of an organization's material goods or services may be subject to governmental question and control and to client reaction.

The synergetic quality of value exchanges is the cumulative total of all benefits and costs to individual members and to the whole of an organized society or social system. Increasingly, American business and government are concerned with the quality of organizational and economic synergy.

As our American democratic society was and is characterized fully as much by its universal values or ideology as by its short-term economic profit motives, the study of social values in America must consider idealistic human-political values as well as economic-material values. A comprehensive consideration of values must include an analysis and interpretation of psychological and logical values, deference and welfare values, ideological and empirical values. In any and all cases the values are pluralistic and interdependent (systematic).

The research of behavioral psychologists indicates that humans have psychological or deference need-values in addition to the basic utilitarian needs for food, clothing, and shelter. Abraham Maslow (Maslow, 1954, pp. 80–122) believes that human psychological needs include physiological safety, belongingness and love, esteem, self-actualization, and cognitive and esthetic satisfaction. He further states that the situational or environmental demands upon the individual apparently can reverse any generally established hierarchy or order of these universal value classes.

Stuart Dodd (Dodd, p. 647) proposes that social institutions traditional to all organized societies represent a consensus of proven or normative human values. He suggests that all complex societies contain domestic, scholastic, economic, political, religious, philanthropic, hygienic, recreational, artistic, scientific, linguistic, and military organizations. The value needs met by these organizations are representative of standardized value ideologies.

Guy Duncan (Duncan, pp. 503–507) suggests that the following human values are rather constant or representative of our American society: utility, distinction, accommodation, continuance, mastery, pursuit, mobility, stability, esthetics, and virtue.

Jacob Getzels (Getzels, pp. 146–161) believes that the values historically representative of the American people are democracy, individualism, equality, and human perfectability. He asserts that a set of secondary social values also traditional to America includes the work-success ethic, future-time orientation, independence, and Puritan morality. Getzels suggests that these lesser values are being modified toward a new value set which includes sociability, present-time orientation, group conformity, and moral relativism.

American educators have attempted to identify social and educational ideals and objectives since the days of Herbert Spencer. Many of the educational value taxonomies are well known. The two most comprehensive and systematic taxonomies of educational objectives by Bloom, Krathwohl, and their associates have been mentioned in an earlier part of this book. In very general terms American educational values reflect our democratic ideology and utopia. We profess a belief in equal educational opportunity, universal education and literacy, and encourage positive attitudes and actions supportive of freedom of belief and purpose, thought and communication, and personal and social choice, action, and mobility.

Although we have long since abandoned the categorical identification and fixed structuring or ordering of values in our American society, we have developed instrumental criteria by which we can more representatively and freely interpret or systematize human values. These systematic valuing criteria include:

1. Universality—a value that affects all men and societies rather than some is of higher priority, other things being equal;
2. Permanence—a value that endures over time is superior to a value of short tenure;
3. Cognitive consistency—a value that harmonizes with numerous

systems of beliefs and credibilities is superior to values that generate conflict among systems;

4. Relative utility—a value that is useful in a wide variety of stable and/or changing situations or one that is crucial to survival is of greater order or priority than values of lesser functional utility.

In more objective, precise, and determinate evaluation or measurement, systematic conditions or properties may be identified which assist in the interpretation of these value relationships:

1. There may be an inconsistency in the part-whole relationships. The scale units may not vary directly with the total value. Unit objectives may not relate to higher level objectives.
2. Indices measuring differences among total values including gain and cost values may include scale units that are not consistent or compatible. That is, the value increment or gain may not be measured in the same units useful in measuring the original value state.
3. There may be too many indeterminate value variables which are ignored because of lack of useful criteria.
4. Only gains may be considered while costs may be ignored. Complete value measurement must consider both total value and total cost and value gain and gain cost.
5. Different and unrelated scales may be used to measure value gains and value costs.
6. The major value to be measured may be only an intermediate or mediating value of a more significant objective, and it may not be a necessary or sufficient condition of this objective.
7. There may be serious errors or omissions in the units or concepts of cost.
8. Alternate criteria of gain, loss, or initial measurement may be overlooked.
9. The time dimension may be ignored in assessing efficiency in the gain-cost ratio or index.
10. Major errors, uncertainties, and inconsistencies may exist in the original value entities, in the instruments or scales of measurement, or in their independent and/or correlated perception.

It should be apparent to the reader why system analysts refer to social systems as extremely complex cybernetic systems that are only quasi-rational in character. Social systems cannot be reduced to determinate or absolute measures.

Systematic research into organizational behavior has made important contributions to the understanding of human values and their functions. Among the first discoveries of organizational researchers was the realization that there are always multiple goals and multiple levels or hierarchies of goals functioning in organizations.

Recent research in the study of organizational values indicates that the maximization of short-range goals or profits and the overly rationalized or standardized focus on money as the only medium of exchange in a social system does not produce maximum overall gains. In fact such a focus is more than likely to conflict with maximum organizational economy, profitability, and synergy.

Computer simulations of multi-level, multi-goal organizations reveal that organizational structures and designs (systems) can be developed to facilitate value-gain efficiency in complex systems, however. System analysts and engineers are applying economic simulators to utilitarian organizational operations with considerable success. History is the best indicator of the success of social-political value systems.

Because American educators have perceived a weakening or changing of some of the more traditional social values in our society, they have tended to accept the inevitability of an educational pattern of increasing utilitarianism, technology, relativism, and materialism. Some educators have generally discounted idealistic, long-range, or abstract social values of any kind and have substituted in their philosophies the particularistic values of systematic utilitarianism, pragmatism, and immediate gain. They focus almost exclusively upon mediating, technological, or process value goals at the expense of any consideration of historically tested and confirmed human and social ideologies and utopias. This restriction of value alternatives and increments often precedes or accompanies a state of anomie or a state of coercive social ideology in government. It is the type of prematurely closed rationalism or pseudo-science that general system philosophers and analysts fear. If too narrowly conceived, an overwhelming national utilitarianism might destroy our nation's intrinsic ideology and purpose. The result could be a retreat from our magnificence and significance as a nation, as the philosophers say. Applied to personal aspiration, it would tend to reduce the significance of life itself.

The general system philosophy of values requires and supplies an eclectic and open human value perspective. It requires self-initiation as a condition of intrinsic valuing and synergy as the basis of a healthy society. A balance among personal and social values and among individual freedoms and social responsibilities is necessary. Both the art and the science of education and leadership require this integrating translation of America's system of values.

It seems sensible to use Ashley Montagu's definition of human goodness or love as a synonym for social synergy. He defines goodness or love as behavior calculated to confer survival benefits upon others in a creatively enlarging manner. (Montagu, p. 11) It is difficult to think of a better definition of social synergy.

Immanuel Kant expressed the same realization in his categorical imperative which states that one should act only in ways which reflect his inclination that the basis of action becomes a universal principle.

In recapitulation, it seems appropriate to reemphasize that human values are functional constraints on human behavior which are created or discovered in order to give both significance and meaning to life. Human values establish man's humanity. They are man's means and ends, ordering all of his social and ecological relationships.

15

Decision-Making in Organizations

A decision or judgment in a human cybernetic control system corresponds to a determinate action in the physical domain of the individual. It establishes new conceptual boundaries which reorder all values or variables. It reorders the perceptual-conceptual universe by reordering the input-output value hierarchy and all relationships within it. A decision effects a closure of the system in some respect and an opening in some other respect. A relatively irreversible transformation of a cybernetic control system is effected when a determinate or binding decision or value commitment is made.

The qualitative aspects of a decision are its unique and/or consistent characteristics, the belief and conviction or credibility which accompany it, and its appropriateness within the particular situational context and within the larger boundaries of the system universe. The quantitative aspects of a decision event are the degrees or scaled range of its influence of effect and the amount and rate of change that precedes, accompanies, or follows it.

If the subjective credibility or belief in a decision is high and its probability of actualization is low there is likely to be considerable emotional tension and uncertainty accompanying the decision event. The reverse is also true. The easily assimilated decision is one where the possibility and probability are both perceived to be high, where the reality and actualization indices indicate its security and economy (its value).

Decisions by individuals require translation in order to be communicated to other members of a group or organization. Originally beginning as personal beliefs based upon perceptions of imagined realities, they become personally acknowledged when they are actualized or experienced in some material way and then compared to alternate or prior experiences. They become public information only when they can be demonstrated or communicated, when their belief values and reality benefit-costs are mutually known and shared.

A personal decision becomes determinate in meaning and in fact when it is understood, personally initiated, implemented, and controlled, and when it is too costly for the individual or his environment to reverse the decision fact. The same conditions define the determinate characteristics of a social decision. A social decision is determinate when it is understood, deliberately initiated, implemented, and controlled by the entire organizational system and when it is too costly for the organization and/or its clients or environment to reverse it.

Most individual decisions are immediate and particular in character. In such a situation there is an almost simultaneous assessment of goals and probabilities, the activation of an operation to effect the new goals, and the determination and evaluation of the effect of the decision. They involve minor changes in the internal value system of cognitive code of the decision-maker and in his credibility system. They are frequently alterable or reversible. Personal decisions of this type often involve strategies of intuition and forecast. Systematic calculation and prediction are hardly necessary or desirable in making many unimportant human decisions.

Social or organizational decisions involve a larger, more complex system, however. They affect a greater social and material domain and probably affect various human elements in the system differently. Their domain and range are more variable and uncertain, as are their benefits and costs. It is likely that the time needed for establishing an organizational decision's a priori belief credibility, its actualization, and its a posteriori evaluation or judgment will be much greater. It is also very unlikely that extensive agreement about either the input or output values of the decision ever will be reached.

Students of decision-making in organizations have ascertained different classes or types of decisions. Decisions may be inspirational or highly subjective, judgmental and/or compromising (semi-objective), or highly objective, computational, and analytic. These categories of decision credibility and certainty are illustrated in the model of Fig. 12.

In a closed or determinate decision-making situation, goal preferences and sure means for their attainment produce functional certainty. In this class of decisions there is no genuine personal choice among alternatives possible or necessary. Decisions here involve straightforward analysis or computation. These decisions involve rationalized certainty or determinate meaning. Operations research techniques involving math-

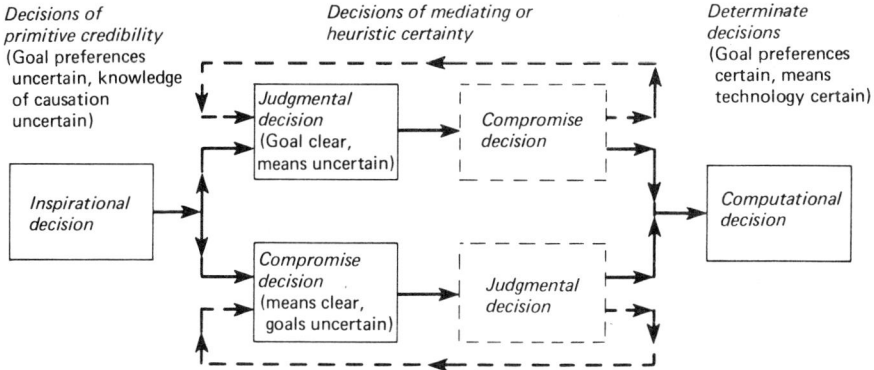

FIG. 12. Credibility of decisions.

ematical and statistical decision-making procedures are designed to facilitate this type of analytic and computational decision-making.

When causation is uncertain or disputed but goal preferences are clear, the decision required is a judgmental one. In judgmental decisions there is considerable uncertainty concerning means possibilities and probabilities. Decisions of this class might reflect disequilibrium between conceptualized goals and material possibilities. Judgmental decisions are necessary in situations where there is a preference and agreement concerning the imagined goals or benefits which exceed the technological and cost certainty of the means. Excessive reliance on this type of decision-making may stereotype or oversimplify decision patterns. Such a strategy of decision-making could lead to the pursuit of imagined but impossible goals.

Obviously the systematic conversion of a judgmental decision into a calculative one is fairly simple. It requires primarily an exercise of skill in the technical construction and demonstration of prototype models which assure technological workability and the economical design, production, and operation of one or more working models. The resolution of judgmental decisions is primarily a problem of science and system analysis.

Situations in which there are ambiguous goal credibilities (multiple and/or conflicting goals) but where there are certainties (preferences and capabilities) of means require decisions of compromise. Here the quantitative dimensions or immediate probabilities, the workability and cost-economy, are given preference over their qualitative elements, their long-range social benefits or values. Compromising decisions seek to accommodate competing preferences by reinforcing only short-range or mediating value goals. This type of decision trade-off is exercised in situations where utilitarian benefits are competitive, immediate, particular, personal, and unequally distributed. Too great a use of compromising or bargaining decisions also will generate error or inconsistency in a decision system. There will be an overcomplication of decision relations. Eventually utilitarian compromise tends to effect a marked disequilibrium between long-range value goals and long-range material means.

The problem of converting compromise or bargaining decisions into calculative decisions is a political-social problem and extremely complex. It involves the building of shared beliefs, faith, and trust in the wisdom of the organizational perspective—its ideology and utopian vision. Representative leadership in addition to technical leadership is required in the productive resolution of compromise or consensual decisions.

The class of decisions characterized by extremely low credibilities and consequent uncertainty of both means and ends is the class of inspirational decisions. An inspirational decision involves intuitive creation or discovery. An inspirational decision is significant, primitive, and unique in relation to other decision consistencies. It begins a series of decision events or a decision chain which must take it progressively through the judgmental and compromising stages before it can be impartially assimilated or calculated.

All classes of decisions just described may be transformed into elements in a philosophical-theoretical inquiry system. The classes or states of meaning consistent with the patterns of decision-making just described can be illustrated in the manner of Fig. 13.

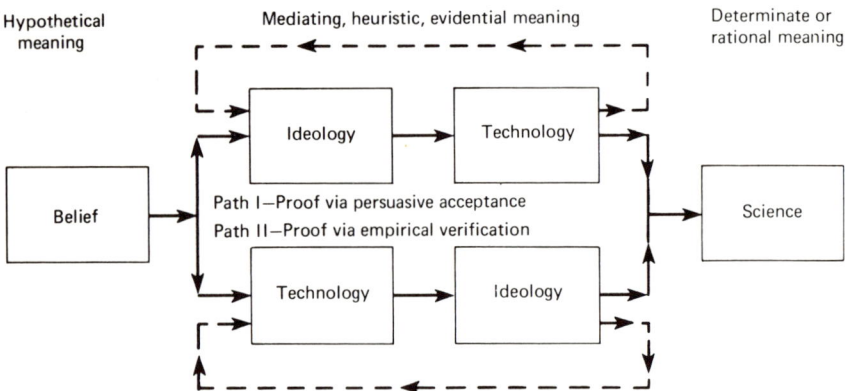

FIG. 13. Decision-meaning classes.

A complete rationalization of individual decision events requires the movement from individual beliefs and inspirations along one of two alternate paths: a path involving collective political-social idealization and normative judgments or one involving extrinsic, compromising, technological patterns of preferences and means of verification. Ultimately systematic means-goals preferences may be verified by computation and scientific method only if the rationality of both the goal and technological systems permit this degree of closure. Truly scientific decision-making as reflected in calculative operations involves a "perfect" resolution of both the ideological and technological elements in the system. The dynamic nature of human individuals and societies seldom permits anything like this degree of perfect enumeration or rationality. Operationally, only a pseudo-ideological and pseudo-technological "perfecting" technique must be used to arbitrarily or mutually fix tentative rather than determinate ends and means.

DECISION-MAKING IN SYSTEM ANALYSIS

The MARS model presented earlier in this book and the material on operations research is relevant to any educational application of system analysis and design. All system procedures demonstrate routines for productively resolving decisions in order to increase their validity, reliability, and economy.

Intuitive or inspirational decision-making involves a "happy accident" level of predictive precision or risk. Intuitive decisions may be correct, but they can also demonstrate the precision and reliability of the medicine man and soothsayer. In order to reach the level of precision attained by the "cheerful robot" or mechanical decision-maker, man has to exercise all of the preliminary steps or rules of system analysis and program design presented in the MARS model. Then he has to behave like a machine, automatically calculating the results of decisions and making a rational choice, one entirely dependent upon prior calculations.

All decision-making, including systematic decision-making, originates as a happy accident or intuition, however. Some human being must originate a nonexistent goal as the first inspiration; this frequently involves the generation of an ideal or better future and perhaps a tentative recommendation for a path to that future. Yet inspiration which is never acted upon always remains at the precision level of the improbable impossible. Creative decisions must be communicated to others if they require organizational or social support for their implementation.

In the communication of beliefs or inspirational decision, some degree of rationality is induced into the decision chain. If an individual verifies his beliefs personally, he must initiate action which produces observable results and then make a judgment about the act and also about the value of the action. This advances ideas into the level of technological compromise with reality. In this manner an inspiration is advanced into the domain of the improbable possible. If an individual communicates his beliefs and tests them by shared judgment only without reality testing he advances his decision-making into the domain of the probable impossible.

It is in the nature of human decision-making to partially answer or assume an answer to the "who" or subjective question in the most primitive presystematic decision-making chain. Also the "why" or output goal or purpose question is partially answered or assumed in the most primitive decision and perhaps the "what" or input value question is partially answered or determined during the subjective creation of an inspirational decision.

The first resolution of inspirational decisions begins with the objectification of the "who, why, and what" elements. Primitive analysis and control require either the conceptualized reconstruction of an ideal or image or the methodological-technological reconstruction of a causative event. Determinate meaning or understanding of a cause-effect relationship requires that it be confined both conceptually and materially and affirmed as valuable by the same process. It is important to note that the order of sequence in the objectification of means or goals is initially arbitrary. Ideals or ideas may precede prototype artifacts or vice versa The perfect "happy accident" or inspired decision occurs when the act and its cognized value originate simultaneously.

Personal decisions become publicly communicated or intelligent when the purposes, methods, and values of causal events are experienced and shared. The ascertaining of complex and ill-defined goals and means (analysis and judgment of the meaning of actions) are communicated primarily through the media of history and art. The method of science decomposes or analyzes act means-ends relationships or operations to ascertain their immediate or particular meaning.

The first step in resolving subjective decisions into more meaningful decisions requires the objective definition of purposes or objectives and the determination of necessary and sufficient input means for their attainment. An early step in complex organizational decision making involves the establishing of a consensus or agreement on the goals and means, one that persists until the decisions are actualized. Because complex technology can be very expensive, it is usually desirable in organizational decision making that organizational purposing or planning and consensus development precede technological design in the process of planning and resolving complex chains of events and decisions. That is why the process of system analysis begins with the goal setting and analysis phase and then progresses to the design and implementation stage. This progression insures the basic integral perspective and economy inherent in the system process.

Following the scientific method of controlled induction in which purposes are tentatively established by consensus, organizational decision-making progresses to the analysis stage. If we persist in the assumption that conceptual modeling is generally more economical than technological modeling, particularly when choices among alternatives are to be considered, we will conduct our logical analysis of statements of purpose, analysis of relevant research, and at least our internal political analysis prior to or simultaneous with our technological analysis and economic analysis.

As indicated in the MARS model, all of these types of conceptual analysis must be completed in full-scale system analysis or planning. The order in which they are done is somewhat flexible, depending primarily on the degree of uncertainty and/or cruciality involved in each analysis. Ordinarily, if there is considerable prior control and certainty of output goals and input resources for planning and research, the several analysis procedures (particularly the more complex, critical, and expensive political or market analysis and the technological analysis procedures) will progress simultaneously.

System analysis, as previously indicated in the MARS model, involves simultaneous or alternate interdisciplinary analysis cycles or nested loops which must eventually be integrated and coordinated before a single culminating decision selecting one from among several plans or models can be attempted. Ordinarily the final analysis made prior to making a choice among plans involves economic analysis and the calculation of the most economical means to implement the plans.

In system analysis objectives are established first and held constant; logical and theoretical confirmation of goals and possible means comes next; assurance of technological workability and the enumeration of several workable methods and means follow. Then and only then can the final decision problem of calculating input costs and market or product benefits be undertaken. If all prior decisions have been made carefully, the calculation of comparative benefits and costs ordinarily automatically determines choices. Consensus and judgment are almost totally assumed (built into the system) at this point. In the calculated decision the system or technique prevails over human judgment, which is not fully necessary or functional. Thus if a process of system analysis is thorough and complete, the techniques of operations research can be applied to most or all of the culminating critical decision elements. In ill-defined or extremely complex problems most of the calculated decisions are based on estimates, however, and, as pointed out earlier, many problems are not amenable to comprehensive mathematical calculation independent of human intervention by means of compromise and judgment.

In many complex organizational problems, decision-making almost always involves gross estimates based on limited experiences and/or uncertain predictabilities. Predicting natural and technological factors is relatively easy if they are representative of input resources under present control. Predicting the necessary and sufficient economic and political resources under present control or commitment is slightly more difficult. The most uncertain problem of all is predicting the technological factors and particularly the psychological behavior of the uncontrolled social environment. Employees having limited commitments to organizational changes are fairly unpredictable and the uncontrolled client system or market is extremely unpredictable. Especially if either the employee or client subsystems are basically oriented toward systematic counter-strategies of competition is there great uncertainty in estimating system components mathematically.

As mathematical decision-making is applied in the presimulation of any complex system process, it can vary from the highly precise to the highly imprecise. Its quality rests entirely in the quality of the information upon which it is based. Like computer operations, mathematical models are veridically transforming. "Garbage in, garbage out" is the appropriate interpretation of such models.

All precise estimates and predictions must be based upon considerable prior experience and information and upon consistent or normative performance in the long run. They generally assume the additional expectation of human rationality. And they must therefore assume that the operation will in fact take place. Where relatively independent employees and clients are involved, this last assumption is always highly uncertain. It is a political-psychological problem requiring the fixing of beliefs and acts through persuasion and commitment which continues up to the point where the final or penultimate event is transacted and completed irreversibly. A decision chain is finalized only when it is too costly or impossible to reverse the penultimate or final operational decision.

In system analysis the problems of "where" and "when" to execute an operation are rather simple problems and involve simple decision chains. Economically, the "where" problem is usually determined by considering the adequacy of present facilities and resources and the possibility of alternatives which will be less costly. The "when" problem is economically resolved by determining the latest date for starting an operation and commiting resources which is fully compatible with the earliest possible or the optimally scheduled completion date.

The "how much" problem in economical operations research or management science is the problem of maximizing profit. It requires estimating the future market of the product or service outputs and ascertaining the optimum number and mix of input resources. Full technological control of production is coordinated with estimates of production and the desired system operations are complete and subject to reassessment when this state of equilibrium is attained.

In the overall strategy of system analysis the conservative or practical order of progress is constantly maintained following the initially accepted program modification. The strategy begins with simple wholes, progresses to production and economic analysis, and then develops a series of orderly increments in the design and implementation process. This economical procedure so typical of science and system is reversed or enriched only by establishing rather ideal output goals prior to their economic comparison and budgetary restriction. Without liberal amounts of political-social persuasion (advertising), heuristic prototype building or experimentation is seldom justified in carefully applied systems technologies, particularly in the short run. The conservative or evolutionary approach to change is endemic in the method of science and systems. A more liberal approach can only be accomplished through the exercise of the methods revealed in the history and art of politics and persuasion.

In order to illustrate the relationship between the major disciplinary forces affecting human perceptions and the dimensions of liberality and conservativism as they affect societies functioning within their environmental restrictions, the model in Fig. 14 is presented.

The conservative perspective ←			Press of human imagination	→ The liberal perspective		
The economic fanatic	The agent for stability, historian, artist, technician	The natural scientist— philosopher	An integrated system perspective	The social scientist— philosopher	The change agent, social entrepreneur	The political fanatic
Satiated economic demand						X
				X	X	X
		X	X	X	X	X
	X	X	X	X	X	X
Unsatiated economic power demand ←				Unsatiated political power demand →		
X	X	X	X	X	X	
X	X	X	X	X		Satiated political demand
X	X	X				
X	X					
X						
			Press of nature, reality			

FIG. 14. A model of political-economic valuing, decision-making, and judgment.

In this model the qualitative-quantitative relationships perceived to exist in the normative perspectives and behaviors of groups comprising the conservative or inventive forces in society and those comprising the liberal or creative forces are illustrated. The liberal political fanatic intuits economic-social change for its own sake. He seeks to accelerate man's control over nature in a direct linear manner, paying little heed to consequences. The hoped-for result is personal and human aggrandizement. The social result is disintegration of historical-esthetic-ethical values, multiplication of goals, market overproduction, environmental waste.

The economic-political conservative fanatic seeks stability and status for their own sake. He seeks to keep more longer (to lose less faster). The social result is integration and oversimplification of values and goals, market under-production, wasted human opportunity.

You will note that the system perspective presented in the model is an integrated and balanced mean between the liberal and the conservative, between man and nature. It seems that systematic thought and education should guide men to recognize the challenges confronting them in preserving and conserving as well as exploiting their natural environment and in cooperating and sharing with their fellow men. If this does not occur, they will be compelled by their dynamic and imaginative nature to create and invent means only for destroying natural resources and competing with and dominating their fellows.

But let us return to the strategies men employ in effecting group decisions in social organizations. As the rationality in most social organizations is usually far from complete, some form of adjustment of organizational decisions and actions is usually necessary. Competition is one common social strategy that develops from intermediate decision situations involving judgment and/or compromise. Competition operationally involves the acceptance of a limited degree of disagreement and uncertainty between rivals who are attempting to influence the establishment of preferences, the formulation of goals, and the securing of limited resources for attaining these goals. Pure competition is a direct opposition of social power values, a zero-sum game in which one party wins at the expense of the competitor who loses or is destroyed (this competitor may be the environment). Competition is intensified by the perception of social organizations as closed systems, systems with limited and fixed input resources and strong environmental restrictions. It is further intensified by assuming non-negotiable social value constraints which frequently greatly restrict both the choice of new purposes and alternate means of operation.

Cooperation, compromise, and bargaining are other forms of decision-making strategies involving social power-value exchanges. Complete cooperation requires social consensus of all anticipated goals and means and the justification and verification of both input and output values. Compromise, negotiation, and bargaining are necessary where some form of goal or value displacement (either conceptual or material) occurs. Special forms of social compromise include coalition (cooperation for a limited purpose, to a limited degree, and for a limited time), and cooptation (absorbing resistant elements into the organization by buying them off and employing them). Cooptation is a social strategy which tends to replace rivalry or uncertainty with certainty. It eliminates a rival completely in respect to specific conflicting decision situations.

Coalition is the most intricately linked form of limited cooperation. It essentially requires a continuous computation of degrees or limits of unity and diversity. Such an arrangement places rules and restrictions upon decisions in regard to the establishment of new goals and their means of attainment. It does result in an increased certainty or credibility of both perferences and means, however. Informal planning and evaluating coalitions between organizations and their client or environmental systems are examples of coalitions. The relations between employers and employees in a dynamic social organization is something of a blend of social strategies involving full cooperation, coalition, cooptation, and direct competition.

The essential or systematic characteristic of social cooperation is the disposition of the system to focus its primary effort or tension not on internal personnel components and relations per se (internal competition for limited resources), but on mutually agreed purposes and the means for their optimum attainment. Human effort in cooperative social systems is directed toward shared problem elements. The perception of forces linking subsystem personnel is positive and distributed.

Social competition personalizes and atomizes tension and goal focus. It directs it inward toward subsystem personnel and toward displaced mediating goals. It tends to coerce and frustrate personnel whether or not the tension created will be relevant in reaching the common goals of the organization. The residual effect of excessive competition in an organization is negatively reinforcing; it produces interpersonal and organizational alienation, a factor common in America today.

Cooperation requires, reflects, and perpetuates a balance between self-interest and self-less interest. Competition maximizes self-interest and systematically deemphasizes self-less interest. Normative institutions such as schools or governments which seek to develop cooperative ideologies and strategies in societies will not accomplish this by imitating utilitarian business competitive practices.

Computation as an organizational strategy of decision-making is found only where the system can be categorically ordered and closed. It involves definitive means-end linkages and relatively fixed relationships among all elements in the network. It requires total integration throughout the system. Needless to say, computation as a form of decision-making in a complex and dynamic social organization requires extensive prior structuring of human beliefs and assumptions as well as agreement on operating conditions. Surprisingly, calculated or fully rationalized decisions which apply to significant and persisting organizational purposes can only be maintained through voluntary and self-initiated co-operation and commitment to common ideologies. This is the province of greatest interest to philosophers, theorists, and system analysts, who generally recognize the importance of self-initiation and self-motivation in all human affairs and ascribe to a Theory Y philosophy of organizational leadership.

The Theory X leader relies on external motivation, rewards, and sanctions to gain limited commitment from employees in a perceived struggle for limited resources in a competitive environment. He uses his operations research calculative techniques to make decisions which, to no one's surprise, apply accurately only to extremely limited and determinate problems. As competition and alienation increase in a social system, precise calculation of decisions will apply only to fewer and more limited operations. Negotiations, bargaining, and compromise among client and employee subsystems will be more frequent and reflect more self-interests. This is the nature of utilitarianism and materialism.

Obviously any social organization which attempts to remain a productive and integrated system over time must achieve some appropriate balance between cooperation and competition if it is to persist and maintain itself. It would seem that organizational theorists should take another look at the possible necessity of multiple types of organizations in complex social systems (normative, utilitarian, and coercive organizations), rather than attempt to homogenize them into some single standardized utilitarian mold. This need seems to apply particularly to commonweal and professional institutions. We cannot and should not plan, operate, or evaluate in normative or in coercive organizations in the same standardized and limited way we do in utilitarian organizations. As always in human and natural affairs, pluralism maintains and rationality is incomplete.

In administrative decision-making in complex social organizations it should be obvious that all four classes of decision-making (cooperation, coalition, compromise, and competition) will be exercised in an almost infinite variety of ways. The competent administrator will learn to understand the alternative decision options available to him in a particular situation or class of situations and the consequences of these alternatives. The educational practitioner should have no difficulty in concluding that computational decision-making of the kind required to calculate determinate decisions or program a computer is inapplicable to many leadership decision situations. However, the intelligent leader will seek and use information gained by appropriately precise methods whenever and wherever it is possible and practical.

In developing theories of any kind, including theories of system analysis and operations research, the theorist deliberately produces designs or systems of rules and procedures which are relatively insensitive to or independent of changes in assumptions. In other words they objectify and veridically or truly represent reality only normatively or in the long run. Thus they eliminate as much unsystematic bias or error information as possible. They do assume and reflect systematic bias, however, the continuation of human and natural consistencies. Thus the use of theory and other means of inductive and deductive inquiry will never assure absolute or infallible prediction or control of dynamic or evolving social or natural systems. To infer such a level of control is ridiculous. However, the use of the methods of science, probability, and system will move human decision-making systematically and efficiently from the inspirational or high-risk forecast and control level to levels of much greater prediction and control. This is indeed a remarkable accomplishment, and science and system can be expected to do no more.

The science of mathematical gaming is that part of operations research and mathematics which applies most directly to the making of decisions where two or more persons can actually influence or determine the decision-making strategy and results. Mathematical gaming is to problems of social policy-making and strategy what probability theory is to problems of a natural or technological kind (those involving random or chance error rather than strategic errors or biases). The basic principles of game theory point out the effects of the different assumptions and strategies involved in decisions involving two or more quasi-rational cybernetic human beings. Politicians and social strategists or leaders of any kind will find in this discipline the most thorough and penetrating theoretical analysis of social decision-making. As might be expected, game strategists fully acknowledge the complexity and uncertainty involved in their assumptions and rules of theory building and the great probability of error when their normative theories are applied to rather singular or unique problem situations.

Complex system procedures in extensive use in educational management, whether employed by administrative decision-makers or built into routine mechanized data processing procedures, have practical applicative limits. Managerial economics, accounting, and record keeping are the dominant areas of mechanized system applications in schools. A second area where batch system procedures and technology are employed in school administration is in educational research and experimental simulation. Instructional technology using automated machine-mediated processes of decision-making has proved to be most useful in library data retrieval and in instruction in routine basic skills.

Only if and when large scale educational planning can be fully justified in application will system procedures attain a general purpose rather than a special purpose or batch function. Simulation of a total educational decision system is a remote practical possibility although it may have basic research benefits. It is hard to conceive of a situation where it will be either representatively effective or economically feasible. It will most effectively model the upper limits of an ideal or perfectly rational rather than a realistically rational educational system.

Wherever educational decision-making can be applied to standardized and routinized problems, methods of operations research may prove to be useful and practical. Without doubt there are numerous management routines which can be planned more precisely and economically in educational institutions of any size. However, the results of prematurely institutionalizing or closing systems of decision-making or administration are, as has been pointed out in a number of ways, uniformity, standardization, normalization, and short-term efficiency. Applied to mechanized processes, long-term efficiency may result; applied to human behavior, the overall effect may be much less desirable.

Procedures of complex planning, system analysis and design, are applicable to the resolution of many major decisions involving the improvement of all facets of school operations. The application of system routines such as the MARS model can prove to be extremely optimizing in productively transforming or improving present educational operations. A simplified application of system procedures may resolve other less important but equally ill-defined decision situations or problems for the educational administrator much more productively and economically than present more intuitive methods.

A general system philosophy or perspective consistent with cybernetic psychology seems to be a useful frame of reference for approaching all educational decisions. Organizational rationality, conceived even minimally, involves an orientation toward specified operational objectives. It must therefore define expected employee performance and establish justifiable and efficient reward-control systems consistent with these expectations if it is to maintain its limited rationality. Such a degree of systematization requires integration of both the perceptive and cognitive control elements in the system. The unity of command is attained only by an appropriate system of decentralization and distribution of both decision-making and information which are then linked in a network which can accommodate all self-initiated communication and consensus. In modern professional organizations rationality via information distribution and shared decision-making has to a great extent replaced rationality via the limiting and centralization of decision-making. All presently acceptable operational models of organizations tend to indicate some balance among centralization-decentralization forces.

History seems to indicate that semi-rational organizations are more efficient and have a more pervasive influence than more highly rational social units. The long-range trend in organizations has been away from autocratic-centralized social systems characterized by hereditary social stratification, set political succession, complex hierarchies of power centered in small elites, undistributed and oversimplified values, closely controlled reward-exchange systems, elaborate systems of performance evaluation, the tight control of information, and all other forms of social domination and slavery. Thus we are seeing a demonstration of the perspective of the general system philosopher. Social systems are indeed multidynamic transforming cybernetic systems.

System theorists and analysts believe that the optimum condition of interpersonal perceptions in an organization is a state of organizational synergy and serendipity, a state of evolving and cumulative profits and economies. Cooperation is the synergistic mode of interpersonal relations in an organization in such a state and sources and destinations of power, value, and decision acts are multiple and distributed. Some type of shared belief, ideology, utopian vision, trust, and loyalty is needed to integrate the optimum organizational network. These shared assumptions serve as a common base for generating human and organizational purposes and preferences.

Meaningful communication is the most efficient source of employee and client control in healthy social organizations and institutions. The shared determination of social goals is necessary to obtain optimum participation of social inputs leading to their attainment.

Leadership in synergistic social organizations will be balanced, professional, flexible, and adaptable, with the formal leader exhibiting receptive and contemplative behavior along with his self-initiating activities. Communications networks will emanate from all employees and clients. Supervision will be minimal, directed primarily toward improving goal definitions and accomplishments. Supervisors will employ positive persuasion when necessary while, at the same time, maximizing their employees' self-motivation.

Goal displacement will be minimal; employee identification with organizational and social goals, maximum. Work attitudes will reflect and maintain whatever intrinsic perceptual values are possible. Therefore, commitment will be personal as well as contractual and leadership will be positive, cooperative, and effective to the maximum degree.

In this open and dynamic social system a mixed (alternating) strategy of activities will improve on the normative strategy of the closed organization. The system will be open to change from within as well as change from without, and, because of this, employee satisfaction, loyalty, and morale will augment total material or economic dependence as a factor influencing employee retention and support.

If the reader feels that the interdisciplinary system perspective of administrative leadership as presented here is idealistic, he is entirely correct. There is absolutely no idealism or long-range motivation and growth in a closed system based upon an erroneous or mythical belief in the existence and desirability of pure human rationality. Human wisdom is more than mere human intelligence.

Yet it should be strongly emphasized that a systematic perspective of administration or education and the deduced strategies of system analysis and design are not naively idealistic or simplistic. They systematically accommodate both the imaginative, idealistic, and uncertain in human behavior as well as the informative, realistic, and certain. It thus accommodates the power of human and organizational synergy and serendipity which can be diminished in the misapplication of more highly rational management science strategies.

16

The American School—a Social Institution in Transition

As America enters into the decade of the seventies, it is quite apparent to all that its system of education is undergoing a marked transformation. In such a period of obvious transition it is difficult to predict the outcome of present plans, experiments, and innovations. Perhaps the best that can be done is to attempt to identify and assess the primary changes that are taking place and to speculate on their implications for future educational developments. A system perspective and the interdisciplinary theories of organizational leadership just presented are employed in the formulation of the following paragraphs devoted toward this end.

Schools as organizations are social subsystems. They are units within larger social units. They, in turn, are composed of semi-independent, semi-dependent subunits (employees, student clients, parent clients, and occasional other community environmental subgroups). And they function in a social universe containing numerous other types of relatively independent organizational units.

As socializing commonweal institutions, schools in America have served fundamentally as normative institutions. The properties that have persistently identified the American public school among all social organizations are its democratic ideology and its utopian vision of human perfectibility. The major mission of the schools is comprehensive education, the indirect operationalization of America's consensual yet pluralistic ideology and utopia, its present and emerging social, political, ethical, and economic goals and processes, within a dynamic and integrated program of learning and living.

Schools, as socializing yet professional organizations, offer their clients primarily one product, a service. Yet this service is becoming increasingly complex, as is our society. During this century the schools have been severely pressed to extend their mission or service to include increased vocational and prevocational training, numerous health and welfare functions, and their general custodial care of a greater portion of American children and youth for a longer portion of their lives. At the same time they have assumed a greater role in expanding their normative function of transmitting the culture and the accumulating knowledge of our society.

Although discounted by many professional educators, the custodial service of the schools is a well-established and expensive part of the total educational service function, and it has markedly increased during this century. (Educators generally do not mind serving as the custodians of social culture, nor have they resisted in general their increasing responsibilities for vocational and welfare services, but they are not inclined to stress their institutionalized responsibilities for the physical and disciplinary custody or "baby-sitting" of American children and youth.)

Thus we find that during the present century our schools have greatly expanded all of their educational service programs, their program of cultural transmission, their vocational and welfare services, and their custodial services. In some cases they offer free or low-cost public education from the preschool through the college and graduate school. At the same time, due to the changing social practices and culture of our democracy, they have voluntarily and in response to enlightened public demand decreased their coercive leadership and custodial character and function. Yes, the schools of today are vastly different from schools at the beginning of the century, whether or not they have changed their instructional content, techniques, and materials. And it is probable that the instructional changes occurring during the last few decades are the most remarkable and persistent educational changes that have taken place.

The scope of the services that schools are now offering their client communities characterizes them as the most comprehensive normalizing institution in our present culture. In addition, in our generally pragmatic and utilitarian culture, the American public school is perhaps more idealistic and future-time oriented than any other major social institution. Therefore, it should surprise no one that schools are becoming the focal battleground between (1) social forces in our nation that see the defects and limitations of our past and present ideologies and practices and seek to implement a new and better utopia, and (2) those forces that seek fearfully and blindly or deliberately and knowingly to maintain the status quo.

As to the professed dilemma of whether the schools should transmit the culture or serve as an instrument for cultural change, the issue is really a psuedo-issue. Intelligent people know that any entity which has the capacity to guide or control human behavior, information, or other material and natural resources will accidentally or purposefully transform them. Schools cannot avoid transforming their external and internal social environment, nor can they avoid transmitting the social culture of their encompassing social environment throughout their internal structure.

The question of leadership for cultural change in American education then becomes, "How can schools improve themselves and their society most effectively and efficiently?"

Traditionally, cultural transmission has been more common outside the school than within it, for, even in America, the truly comprehensive and dynamic school has been very slow to emerge. It has not really emerged as yet in many communities, although knowledge of its existence is rather widely disseminated. Therefore, even if we expect that the comprehensive and dynamic character of modern inductive inquiry processes will eventually characterize and shape education in the future, it appears that the last few decades in American education reflect a final short-range trend toward a traditional mediating utilitarianism and a fragmentation of educational curriculum as the basic characteristic of the period.

In support of the thesis that traditional political and economic utilitarianism has been the most pervasive force of mass education in recent years, this evidence is offered. From all levels of government and from all other social institutions in our society, schools have received requests and demands to stress the utilitarian vocational function of their educational programs. Work centered education is expected by many to necessarily encompass a longer span of each individual citizen's life.

Although advanced vocational education is becoming so technical that it is currently placed in the junior college, technical school, college, or professional school, a greater portion of both the elementary and secondary school's resources are devoted to education oriented toward prevocational training. The cognitive-scientific-systematic subjects are being stressed at all levels of the so-called general education program. And, although these subjects may have certain humanistic or esthetic qualities, this is not the primary reason for their justification nor is it often reflected in their teaching.

It is important to note that the major academic curriculum changes which have received the widest dissemination in the past decade in American education have been in the academic areas of pragmatic communications, mathematics, science, and in the vocational subjects. Material and technological resources necessary for major changes in these mediating or utilitarian areas have been the first provided by the federal government following earlier vocational education grants. The value of these changes seems to be based in a broad consensus supported by government, business and other social organizations, and by parents and educators alike. They have not been perceived to be threats to the generally recognized ideological and political values held by our society. These changes have been congruent with the traditional pragmatic and utilitarian philosophy of this nation.

The knowledge explosion and advancing technological developments within our nation have accentuated the utilitarian function of the schools. A political-economic-technological power gap between technical and non-technical communities and nations is readily reflected in all of the common economic indices and political news which is so widely communicated in the modern world.

It is really not surprising that vocational-technological emphasis is the pervasive educational force that it presently is. Since World War II and Sputnik, economic and political competitive pressures upon American society and its schools have been tremendous. International competition and coercion have dominated the world scene. In addition, large segments of present-day society, economically and socially displaced by automation and the effects of technology and further displaced by changing educational and psychological-philosophical needs, are struggling desperately to maintain their economic-technological advantage and/or status quo. Improved world and national communications is more effectively disseminating the failures of past societies, including our own, and the resultant truths are themselves a threat to the insecure, the ignorant, and the uninformed. And, of course, some political demagogues emerge to exploit the social threats and pressures of any technological change to their particular advantage. As a result, large segments of our population—many of the elderly, the tradition-dependent or ritualistic, the conservative, and/or the selfish successful—are actively or reactively resisting further educational and social changes, particularly misunderstood ideological changes. They struggle to preserve the goal-displaced educational and institutional values of the past, often clinging to now empty social forms and conventions, whether or not these elements have any esthetic, ethical, or functional utility.

But the great masses of our society, as always, readily adapt to their environment. They try to play the pragmatic game so characteristic of our culture. The pressures on the schools to innovate, to exhibit technological and methodological changes even if they are educationally insignificant, are great.

The most readily observed changes that have taken place in American education during recent decades are the technological changes in school facilities, instruments and tools of instruction and communications, and in those products facilitating the educational service functions of administration, guidance and counseling, health and welfare, and transportation. Instructional facilities, materials, and equipment, including modern communications media such as video-tape and television, have greatly enriched the traditional teaching and learning environment. Better designed texts, programed texts, and computer-assisted instruction have also enriched teaching and learning methodologies.

Educational practitioners know, however, that these technological devices have been employed generally to improve traditional educational processes and learning experiences, usually emphasizing traditional concepts. Used in this manner, the new technologies will generally produce only that short-term motivation effect which accompanies any innovative change. As merely an alternate methodology, the long-term benefits of modern technologies lie either in their incremental economic impact or in those few instances where they indirectly or directly facilitate the achievement of particular learning responses and activities which were previously impossible. These longer-term benefits are probably more of an exception than a rule in those cases where the technology has been applied in traditional teaching without changing the teacher's perception of the subject content goals or his relationships with the learner.

New technologies have been applied in administering and organizing students for instruction. There they have also increased options and choices. Alternate methods of pupil grouping and scheduling are feasible. Team teaching and other modes of staff utilization have been made practical or competitive. The above techniques can be used singly or in combination. Together these innovations which were made possible primarily through the advancement of technology, have provided necessary but not sufficient means for changing basic instructional (pupil-teacher) relationships. We can now flexibly schedule students and allow greater individualization of instruction in the traditional educational environment. One should note that these capabilities do not necessarily change teaching practices or learning content, however.

Although the momentum for technological innovation in education appears to be great at present, it is probable that many of the changes based upon technology alone will produce only limited or insignificant effects. They will survive in competition with traditional methods and means or other innovations only to the extent that they prove to be more workable or economical. Technology itself, in the hands of the business entrepreneur, tends to deliberately induce and reflect planned obsolescence. And what men can arrange they can change. Technology and organizational change will tend to assume real importance only to the extent that they can be associated with and justified by systematic theory-guided and philosophically-guided human social processes. Thus most of the innovative changes occurring in education today are doomed to impermanence.

There are some observers who believe that educational organizations, due to combinations of system strategies and the options provided by technology, will change so extensively that the traditional custodial service of the schools (compulsory attendance of students for x number of hours, days, weeks, years, etc.) will be affected. If this occurs the schools as organizations will be transformed to a degree hardly recognizable. Yet, as the cost of custodial care is increasing, and, as the ideological and esthetic missions of the schools in the traditional sense are frequently restricted or deemphasized, this degree of organizational change should be considered a possibility. It is but a matter of simple inference to hypothesize that social economies and perhaps even limited social or political benefits may be effected by the separation or fragmentation of the instructional and custodial functions of the schools.

It is probable also that educators and their public will soon consider the possibility of effecting other educational economies and benefits by separating the recreational and/or enrichment programs in education from the vocational, prevocational, and custodial functions. New groupings of all of these educational functions are readily feasible with modern technology. For example, socialization experiences can be separated from individualized computer-assisted instruction in academic or cognitive skills. There are numerous media and organizational possibilities for guiding learning in all of these areas. If we apply system rules of procedure, we will choose the most workable and economical methodology from among better alternatives for each particular educational task, as long as it does not compromise the immediate or long-range educational needs and benefits of the student and his society.

If we combine technological capabilities with emerging perceptions of greater freedom of choice by individual learners, we have still another set of educational innovative possibilities to consider. We may realize other educational benefits and economies by departing from compulsory attendance and compulsory curriculums; individualized criterion promotion, programing, and grading or reporting may be employed to replace normative patterns of promotion, programing, and grading. All of these techniques or processes can assist either in individualizing traditional or mass-education instructional processes or in guiding modern methods of individual inquiry.

The size and nature of the educational technology market has promted a wide-spread commercial interest in schools. It has stimulated a number of mergers between major computer manufacturers, communications media, and textbook manufacturers. The common interests and capabilities of these information generators and distributors is obviously great. It appears that business corporate interests in schools reflect the expectation that changes occurring in educational organizations and their services can and will provide a very significant commercial market.

While there is little doubt that almost every aspect of traditional education can be improved or reorganized in some way or to some degree and that changes in learning concepts and processes will facilitate even greater changes, it seems likely that one or more social institutions specifically commissioned to carry out the educational mission will continue to exist in our complex society. It is also likely that government will be required to assure the equitable distribution of whatever educational services are provided all members of that society, and that it will establish and maintain a special agency to coordinate this mission at each governmental level and to implement the mission at the level of application. The most important responsibility of the professional educator is to assure that meaningful and comprehensive learning experiences are guaranteed each person as they are needed, desired, and employed productively. Beyond that the educational leader should be open to any and all options which will achieve the educational mission successfully and economically.

All technology is but a facilitating tool. It can be used both to productively economize and conserve and to enrich and transform. And, it can be used destructively. Its values are, of course, dependent upon its human applications. What educators do with modern technology and organizational capabilities is more important in both the short and long run than the application of technology merely for the sake of innovation and activity.

Educators must apply the criteria of economy and feasibility if they are to systematically improve the use of technology in education. They will sometimes find that the ideas behind technology, their systematic meanings, are more important and useful than the technology itself. Techniques of system analysis and operations research are examples of system methodologies which can sharpen the decision-making processes required to make major changes in purpose and process more intelligible and manageable.

The most pervasive changes that have occurred in American education in the last few decades have not been changes directly associated with utilitarian economic gains, vocational and technological advancement. They are related to these mediating and instrumental elements, however. The truly critical and persisting changes which have occurred in education are reflected in the fundamental but subtle changes in the structure of the academic disciplines and in the emerging purposes, communications processes, and learner responses to the new instructional content. These changes are inexorably reflected in the theories and philosophies of scientific and systematic inquiry, in modern mathematics and science as well as in linguistics and semantics and in the social disciplines. These basic changes in academic content and process invoke the learner to think more freely and to inquire. They eventually affect his interdisciplinary intellect and wisdom.

One can forecast the eventual communication of these basic social and ethical value changes throughout our society, particularly since not only academic leaders but leaders in every aspect of life are beginning to see and appreciate their embodiment in technology and interdisciplinary methodology in so many varied circumstances. The theories and philosophies of science and system will emerge inductively from the observation of and reflection upon applied systematic methodology and technology—from the process of systematic inquiry.

The eventual full momentum and import of the technologies and methodologies of science and systems rest in an interdisciplinary system of operating rules or theories and in the induced resolution of these rules and theories into a system of human values or philosophy. A dynamic system philosophy and interdisciplinary system theory must consistently and adaptively reveal the true and functional relationships between particular events, routine or systematic operations, abstracted rules of operation or theories and abstracted systems of values or philosophies. Sooner or later educators will have to acknowledge the necessity of accommodating their historically proven academic and applied disciplines singly and in combination with one another in a truly dynamic, comprehensive, and integrated philosophy of education.

It is probable that American education is beginning to sense and reveal the irreversible effects of our cybercultural revolution, the emergence of an integrated and interdisciplinary systematic philosophy. Modern education is experiencing a period of transition which has social and ethical import as well as technological significance. It is quite obvious that many of America's intellectuals, its most intelligent youth, its most capable minority group leaders, and its most informed international associates see the inconsistencies between traditional American slogans and dogmas and present technological-social performances. The inquiring mind of the future, wherever it may be found, will not tolerate these obvious dichotomies for long. And neither will it tolerate a condition of social-ethical-esthetic anarchism and social anomie for long.

The Ordeal of Change by Eric Hoffer is recommended reading for educators who believe that traditional societies and traditional educational leaders have remained compatible and complementary for very long. Hoffer speaks against the possibility of a meritocracy of education in any society. Yet it is probable that Hoffer does not fully explain the unique character of our present period of society-educational relations.

It is one in which the transcending effect of systematic human inquiry is supported by a tremendous complex of advanced machine and computer technologies. The cybercultural revolution is more than a slogan.

Next let us consider evolving political and social interaction practices currently endemic in our schools. The professionalization and specialization of educational administration and instruction have been important factors in educational affairs during this century, particularly during the last decade. Knowledge and ability in specialized technical subject areas has been in great demand in the colleges and in the job market. Possession of such knowledge has provided a base from which to acquire social and economic power for administrators and teachers. Chronic teacher shortages in certain subject areas or at certain grade levels also have broadened the political security base of some teachers, although many of these shortages appear to be declining. Increased political know-how in the traditional bureaucratic and competitive sense has accompanied the intellectual, political, and economic transformation of the teachers, generating a new social-political force in the educational affairs of the schools.

Simultaneous with the increase in technological innovation, basic changes in learning content and methods, and the professionalization of the teachers and administrators, there has been a general decrease in the acceptance of overt authoritarian power, coercion, paternalism, and influence by institutionalized (top-down) bureaucrats—formalistic boards of education, educational administrators and supervisors. As a subsystem within our evolving democratic society, the school, although overwhelmingly traditional and bureaucratic, has gradually developed a more democratic leadership expectation. A greater social distribution of economic, political, and informational power has facilitated this change. Major efforts to democratize American schools further, to allow even the students some voice in educational decision-making, seem to characterize the current educational milieu.

In general educational administrators have kept pace with the processes of academic, technological, political, and economic changes in education better than they understand the ultimate meaning or significance of the processes. They are presently acquiring new political strategies and skills of persuasion to assist them in maintaining their traditional leadership roles. Educational administrators have been remarkably successful in meeting the realistic demands of increased enrollments; they are currently working hard to accommodate large numbers of educationally and culturally diverse and handicapped students. These practical matters have demanded almost full attention.

It has not been easy for administrators to keep pace with the more subtle changes taking place, changes in academic disciplines, in learning theory, in social, political, and economic philosophy, and in the professional development of teachers. Thus it is to be expected that the professional educational leadership role has been partially assumed by liberally informed teachers, national political leaders, educational scholars, and even by erudite parents and students wherever other situational factors have permitted this development.

Out of the politically dominated atmosphere in our present schools, the power in the profession obviously has moved considerably away from administrators and into the hands of teachers. Some of it will shift in a similar manner to client groups of parents and students as soon as they become better informed and organized. Only through the balancing of power or through the abandonment of power politics in much educational decision-making will the present political atmosphere in the schools be further transformed into a more idealistic, goal-directed, and cooperative state.

It seems appropriate here to restress the order of values assigned by many professional educators, system analysts, and behavioral scientists to the critical components of the educational process. They assert that the purposes of the learner-client and the relevance of the instructional concept and experience inputs communicated to him are the primary and significant conditions of effective education and communications. They believe that a focus on the learner's responses to educational stimuli are much more important than the instructional method or media employed or the administrative organization used.

The relevance of the instructional message, its conceptual and experiential impact or effect is that part of education that is the most crucial. And the determination of the educational input-output transformation is pretty much within the control of the learner. As he matures and learns how to inquire and reflect, his control over learning increases proportionately.

Preceding paragraphs have attempted to show that the forces for change in educational administration during the last two decades have roots in diverse and multiple social and environmental conditions. Educational leadership theory and systematic applied research, together with new methods and technologies of inquiry, communications, and instruction, have influenced the development of a new perspective of educational leadership.

In the final analysis it would seem that American educators will eventually show the way in resolving education's teleological, political, and economic problems and display the leadership expected in an efficient professional institution. Almost inevitably this will be accomplished only through the accentuation of greater freedom of inquiry and a more active pursuit of human perfectibility rather than the restriction of these optimizing requisites. Based in an evolving and persisting idealistic humanism, education will restress the achievement of multiple yet integrated goals, open and systematic inquiry, client orientation, and multiple benefits and rewards.

The ideal professional educator of the future will be an integrated generalist as well as a competent specialist. He will possess rich experience and sound theoretical knowledge in addition to a healthy self-confidence and independence and a sincere human interest and concern. Participatory professional educational leadership will focus on the individual uniqueness of client problems and needs and will work to achieve an optimum balance of service, commonweal, mutual benefit, and management efficiency goals and means. Human interpersonal relationships in future schools will stress equality and interdependence rather than dependence and structured hierarchy. They will be based in a psychology of faith, trust, and confidence rather than one of fear, distrust, and hostility. The professional educational system will be productively open, not bureaucratically closed.

Ultimately informed educational leaders will conclude there is but one systematic educational strategy open to them. They will realize that human beings and societies can and must improve not only their amoral systems of natural science, deductive thought, and standardizing technology, but also their inductive social and esthetic systems of belief, ideology, and ethical human responsibility and concern. The strategy by which they can accomplish this involves only the operationalization of cooperative and systematic purposing, together with the demonstrative and interpretational responsibilities attendant to it.

V

LEARNING AND INSTRUCTION— THE PURPOSEFUL DEVELOPMENT OF HUMAN QUALITIES

17

Personal Inquiry and Meaning

As it is the expressed purpose of this book to relate the system frame of reference to instruction and inquiry, this part is devoted to such a mission. It is expected that the reader will soon rediscover the interdisciplinary nature of the system perspective which makes it particularly applicable as an explanatory analogue or model of that human social process we call education.

A system has been defined as any persisting set of objects and operations. Its primary capability rests in the reconstruction of events, deliberately or accidentally transforming input elements into consistent or standardized output conditions. It is closed or invariant to the degree that it maintains itself and operates in this reconstructive or repetitious manner and open or conditional to the degree that it transforms itself, either its component parts or operations, and its environmental relationships.

Human beings are unique in their systematic properties. Although all living organisms are dynamic or homeokinetic rather than homeostatic, men are the most dynamic, adaptable, and flexible of living systems. They therefore exhibit extremely complex systematic properties of order, variety, and control.

Men have the particularly unique capabilities of altering and/or stablizing their complex cognitive control or thought centers, both before and after the fact of an event or experience. They are unique in their capacity for successive adaptation, for self-initiating and self-regulating their own social interactions and the natural and technological elements and forces they come in contact with. By systematizing their memories and systems of communications, humans can modify cumulatively or incrementally their own thoughts and actions and their patterns of social interaction. And they can accelerate and alter natural and technological forces and events to a degree which continually astounds even themselves.

The capacity to alternately preconceive and reconstruct personal attention, interest, and knowledge as they transform actions and habits is the outstanding characteristic of human beings. Yet, in spite of the unique properties of human imaginative and reconstructive control, even when supplemented by computer memories and calculative abilities, we know that the intellect of man is always limited in reconstructing the logic or knowledge of complex experiences and events. In any single event, the individual-environmental interaction is to some degree a priori in sequence and more complex in operation than the meaning man extracts from it (or usually cares to extract from it). For this reason men must alternately preconceptualize and reflect upon complex events at times and in places remote from direct involvement. This is the only way that they may move from the immediately or directly perceived and known to the implicative, inferential, and abstract levels of understanding in which little by little they acquire more determinate understandings of that which was previously unobservable or unknown. It is interesting to note that cybernetic psychologists define memory as an unobserved generative element rather than a stored reaction. Memory also can be defined as the structure or system representative of a predistributed cognitive-affective control code.

Because of the very real functional limits of human perception and memory, men must selectively identify, stabilize, and simplify their inquiry processes and thought-guided actions. They must learn to differentiate and eventually reorganize only the necessary and sufficient from within the universe of all sense data, objects, and operations capable of perception in any event. This objectification and ordering of values and relationships is requisite for efficiently ascertaining meaning and wisdom. Knowledge of the complex and dynamic natural world is possible only because it is capable of being studied as a set of relatively simple, consistent, and autonomous elements. Differentiated classification and pluralistic analysis (theory building) of relationships among subject-object, object-object, object-operation, cause-effect, part-whole, word-word referent, and concrete-abstract components in systems are essential for comprehending the meaning of any single element mentioned. There is no meaning possible or refinable without establishing relationships of identity and difference among qualities and quantities.

Thus a teleological or selective purposing nature coupled with subsequent interaction and reflective thought are essential parts of the requisite cognitive control conditions for man if he is to acquire new meanings, values, and technologies. In order to act meaningfully and learn from his actions, he must arbitrarily establish or assert hypotheses or product-means relationships before the fact. Then, if he wishes to learn the truth about his ability to determine the future, he must act according to his hypotheses and evaluate the results of his actions. This process of selective purposing when coupled with systematic analysis and controlled induction of process behaviors becomes the scientific method. When enlarged to accommodate other interdisciplinary analyses and controls which determine the optimum political and economic benefit and cost values among possible alternatives, it becomes the method of system analysis.

Men acting in concert or conflict form still a more complex system for analysis. A social system has as its essential components two or more semi-independent cybernetic systems acting in some type of dependent relationship which is further constrained and restricted by a social and material environment. It is at a minimum a double-looped multi-dynamic system in which each healthy individual member can exhibit autocatalytic or self-initiating action and self-controlling reaction.

To complicate the problem of analyzing social systems, humans have perfected quasi-rational structures of inquiry, communication, and technological control which accelerate human abilities to direct both random and irreversible forces of nature in ways subject to their quasi-rational wills. It is therefore understandable that many great minds of the present century continue to recognize the particular difficulties of applying the methods of science to the description, prediction, control, and explanation of human affairs. They realize that the disciplines of philosophy, linguistics, history, psychology, sociology, politics, and economics are only quasi-scientific at best. These are the minds that understand best the theoretical assumptions and methods of science and system.

Fortunately or unfortunately, the explanation and control of social structures, and communication and behavioral codes becomes possible, precise, and practical only by assuming the fundamental postulates of logic, science, and system which simplify via selection and reconstructive standardization the extremely unsystematic behaviors constantly exhibited by individual human beings. The simplification process aids in eventually understanding and controlling behavior within certain ranges or norms of quality and quantity, but it is always retrodictive and does not account for some degree of uniqueness in each event. Thus the prediction and control of human behavior is primitive, complex, uncertain, and imprecise at best. Precise control is highly exhaustive of the resources of the controlling system and the subject controlled. Yet a minimum level of human control through education and communication is the most necessary, profitable, and justifiable component of human effort and inquiry as it

concerns the most critical and crucial resources and/or enemies of man
—man, himself.

Modern democratic social theories recognize and openly accommodate the dual systematic processes of social reconstruction and cumulative modification through education and instruction. Perhaps, more than
any other American educator, John Dewey understood the dual reconstructive and transformative natures of education. He was among the
first systematic philosophers to forecast the irreversible evolution of education and the continual transformation of societies and ecologies. His
fundamental ideas of inquiry are now further explicated in theory and
practice than when he lived; today the public schools of America are
making a solid beginning in the development of programs which embody
Dewey's goals and methodological principles of teaching students to
think, to freely inquire, to learn how to learn.

Dewey's rather oversimplified model of the process of systematic
inquiry is stated below, slightly modified by the employment of system
concepts in the ways they are used throughout this book (derived from
Dewey, *How We Think*, pp. 12–13).

1. An indeterminate situation must exist.
2. The institution of a problem—the problem is sensed and partially
   identified, qualified, and quantified by analysis and symbolic description. Improved definition partially transforms the problem
   into a more determinate situation.
3. The more determinate constituents of the problem are further
   identified. The terms of the problem are stipulated or specified
   further and are perhaps partially related and ordered.
4. Possible relevant solutions or alternatives are suggested. Ideas of
   changes are intuited or invented. This is the phase of significant
   mental induction, imagination, and divergent thinking.
5. The solutions are weighed or compared in a process of ratiocination, comparison, and cost-benefit analysis. Cost analysis includes
   the determination of probable technological success and the resources expended or consumed in the process of achieving the
   prescribed goals.
6. Criteria for evaluation of the alternatives are established. These
   criteria include predefining essential points for monitoring processes or events and predefining the decision points which monitor
   the final product adequacy or output conditions. Criteria for
   evaluation are derived from common sense and experience as well
   as from more precise instruments (logical and mathematical tools
   of analysis and physical measuring devices).
7. Action is implemented or induced and systematically monitored.

8. A posteriori meanings are determined. The coincidence of pre-defined output goals and actual outcomes of the operating system is measured. Interpretation and evaluation of the event is completed.

Israel Scheffler is a current philosopher-educator who elaborates on the human learning or inquiry process also. He describes the acquisition of meaning as the fixing of beliefs. His sample definition of knowing is (Scheffler, p. 21):

X knows that Q if and only if
(1) X believes that Q
(2) X has adequate evidence that Q
(3) Q

This definition asserts that human imagination and conceptualization must eventually correlate external existence with human perception in order to establish any satisfactory level of determinate meaning, consistency, or proof. It asserts also that an individual a priori belief condition or hypothetical awareness must exist prior to ascertaining factual, material, or evidential meaning.

As pointed out in the information analogue described earlier in this book, the meaning men discover in their experiences or perceptions is always partially the meaning they help to put there. Part of the meaning of an event (its message or code) is unique, but part of any meaningful event lies in the code distributed prior to its inception and perception. Thus it is impossible to completely reconstruct, generate, distribute, or communicate all of a particular event or experience. And once an event is directly or fully experienced it alters all subsequent experiences (possibilities and probabilities) in the system. Semanticists also state that once an event is verbalized or reconstructed symbolically it is no longer that event. The acquisition of meaning identifies, defines, records, reflects upon, interprets, and substantially reconstructs events, but it cannot completely regenerate or reverse them.

Scheffler's definition of meaning as the fixing of beliefs requires a series of stages in the fixation process. The initial belief condition involves the development of properties which direct or focus attention and motivation. These properties must persist in directing and motivating an inquirer through subsequent stages of self-initiated search and self-controlled actions by which a believer gathers his evidence or accomplishes his purpose. The initial or hypothetical belief condition must actively sustain an inquirer in his pursuit of a relatively uncertain and perhaps tension-producing course of action. Sustained human inquiry and goal-directed action without personally identified purposing or motivation is not only improbable but practically impossible.

Charles Peirce, noted American philosopher-logician, employed the term *hypothetical inference* to define an assertedly testable initial or a priori belief condition. He defined hypothetical inference as reasoning from the imagined consequent to the antecedent or independent conditions—a time-reversing or recursive process. Other terms employed to define this a priori human-social belief or teleological state are: purposing, goal set, prediction (retrodiction), planning, programing, and value-engineering.

The second stage of fixing beliefs involves a direct interaction of a believer and his objective environment, wherein the transformative event is monitored via his immediate feedback control means of direct observation and manipulation. This level of meaning is called evidential meaning. It involves immediate cognition and simultaneous material translation, a type and level of cognition which permits only limited understanding (the cybernetic black-box phenomenon). The control level possible in evidential meaning involves capturing via immediate recognition and recall or through instruments for recording observations the immediate feedback of the act. Although almost always fairly incomplete, the meanings derived from such evidential experiences are essential for a subsequently more systematic updating of the human cognitive mapping system.

The development of relatively precise and complete evidential meaning states usually involves much repetition of an experience in order to capture enough of the data necessary to reasonably understand or precisely reconstruct the event symbolically or physically. It often requires frequent heuristic or search trials with sequential reductions or eliminations of skill or data errors. It also requires a very well-organized or focused high-speed information recording and monitoring system. The human mind, a cognitive monitor or computer, specifically programed via prior experiences and a designative symbolic cue language, is quite efficient for monitoring ordinary day-to-day human events to the necessary degree of precision if it selects only the minimum essentials from among all of the sensory inputs to control. Evidential meaning is generally inadequate for explaining, interpreting, or reconstructing complex chains of events to any sufficient degree.

The products of evidential inquiry are called facts. Facts are any properties or experiences which can be made sufficiently stable to be counted upon in any further reconstruction of events. Facts begin as pure phenomena or experiences. They become meaningful when they involve an interaction and correlation between direct experiences and cognitive images. Facts may be intrinsic or personal to an individual human, or they may be communicated through shared experiences, thus becoming extrinsic or public. The natural sciences deal with facts which are primarily extrinsic or public. In the social disciplines it is difficult to clearly distinguish fact from a priori belief or unconfirmed opinion in many events.

The evidential level of meaning is transformed into the third or determinate level of meaning, the proving of facts, by reestablishing a material equilibrium, a point of reflection, a concluding checkpoint following an event. Thorough understanding involves alternation or separation between empirical intake and theoretical assimilation. (Mills, p. 74) In reflective thought the inquirer seeks to confirm the correlation of his original conceptualizations and the material facts. He seeks to reconfirm what Alfred Ayer (Ayer, pp. 1–11) defines as the weak verification of logical proof and the strong verification of material proof. This stronger material proof is made public when an inquirer can and does subsequently reconstruct and standardize the essential properties of an event. Thus personal or subjective belief becomes socially acceptable and communicable fact when it can be subjected at will to direct demonstration and observation and finally to measurement and evaluation by multiple impersonal observers. It is then reproducible in different space-time dimensions.

To be properly monitored, the determinate level of meaning should account for all significant data within the parameters of the experienced event and the experiencing system. It must assess and interpret both the act and the action, the totality of all input and output values. This assessment needs to be comprehensive.

Proof is thus the fixation or confirmation of a subjective belief. It requires both the objective fact, consistency, and extension of an experience, and the ultimate significance and intension of an experience. Proofs are definitive codifying imprints which represent the properties of a system as sustained through an operation or event and the gathering of evidence. Proofs are irreversible, distributed, and relatively fixed properties created by reversible or alterable means. The necessary prerequisites to conclusive proof are belief, desire, or purpose conditions; one or more deliberate controlling or mediating operations; and a homogeneous output or product state which can be reflected upon and compared with the predetermined cognitive map and/or other physical reproductions. It should be remembered, however, that a proof is never absolutely complete or perfect. It is merely the closest possible synthetic reproduction via symbol and artifact.

Within the inquiry or belief-fixing process either creation or discovery may precede the other in the order of succession. They are equifinal. In original creation the material object and/or the perception of a truth or existence naturally or accidentally precedes its symbolic mapping or material explication. In discovery the symbolic or logical explication of a proof or belief, initiated intuitively or deliberately, precedes its material recreation. Both creation and discovery are non-systematic prerequisites to more systematic evidential and determinate meaning.

The achieving of a fully determinate level of meaning and judgment in a dynamic human cybernetic system or in multi-dynamic social systems requires considerable interpretation and valuing. Most events are not subjected to this degree of attention and consideration. The careful reflection upon act and action meaning is seldom attempted or attained in everyday living. Careful and complete processes of system analysis and reconstructive design are generally reserved for events which are critical, costly, or crucial to 'ae welfare of the individual or society which supports the process.

18

The Structure and Process of Formal Education

One might think that the system theorist, psycholinguist, and educational psychologist believe that instruction, when based on advanced and near-perfect human knowledge and experience, would become categorically determinate, fail-safe, and capable of complete standardization. This condition would be brought about by an increased capacity for elaboration and detailing of instructional input content, refined technology of process, more carefully sequenced materials to be learned, and more carefully controlled and manipulated external reward or punishment systems affecting the motivation of the learner. This instructional management viewpoint is consistent with a psychology of learning that perceives typical learners as generally purposeless, passive organisms possessing primarily random and/or reversible experiential and imaginative capabilities and desires.

An atomistic or closed system view of instruction must develop a rationale somewhat as follows: A neutral valuing or purposeless learner who is generally passive is conditioned by extrinsic stimulus-response sensory interruption, reward-punishment motivation, and dependency creation. Eventually the learner's natural motivations and understandings are displaced and/or reordered little by little through the conditioning process. Gradually the system of external goal displacement modifies the natural learner, creating a new system of dependencies, perceptions, beliefs, operational skills, and aspirations or value goals. If the instructor persists in his pattern of capturing the learner's attention and continually interrupting and altering his responses, the imposed system of instruction will ultimately prevail.

Eventually a critical point will be reached wherein the learner is so conditioned (the learning is somehow internalized) that he will continue to "rationally" conform to the external system which has shaped him. When this occurs, the controlling acts or cues can be reduced or eliminated.

This type of instructional theory is represented to some degree in the traditional educational thought of the Herbartian structuralists and the associationist psychologists. They generally discount the internal dynamics of the learner and his capacity for cumulative self-transformation. They perceive self-motivation in a learner as a force to be controlled, not a supportive force to be productively exploited.

Many cognitive psychologists, psycholinguists, and system theorists do not disregard or discount the dynamic and independent nature of the human learner to such an extent, however. They suggest that instruction and communication take place between two dynamic and semi-autonomous systems wherein the information source controls information by regulating rate, reliability, redundancy, ambiguity, time-phasing, quantity, and discriminability among inputs. At the same time, the information receiver dominates in the regulation of information exposure, selectivity, capacity, receptivity, and durability. In addition, the direction of communications flow in the learner communication network or in an instructional network varies constantly. They believe that the art of instruction in such a multi-dynamic system is to design a message network and individual messages which positively bind or link the communications source and receiver in the network by mutual interest, intent, and purpose, and by compatible process.

Most educational system theorists recognize the narcotizing dysfunction or oversaturation effect that may occur through misuse of open-loop, interrupting, overdetailed and oversupervised instructional inputs. They are also aware that the too frequent use of external rewards or extrinsic reinforcements is less effective than an optimum scheduling of intermittent rewards which are systematically and successively reduced to decrease the dependence of the learner upon external human resources.

System educators also are cognizant of the fact that, although human imagination is almost infinite in its variety and its dynamics, people learn early to selectively attend, respond, and value. Humans increase their comprehension of complex events by simplification, classification, judging, and deciding. They tend to seek only the necessary and sufficient rather than the universe of data in any event. Overcontrol and overdetail in a learning situation is exhaustive to both the learner and the instructor. It could and probably would continue to exhaust the learner even if the instructor were replaced by technology, as is the aim of computerized and programed instruction. Human individuality cannot be manipulated into a standardized form except at great cost to both the manipulator and the manipulated.

Thoroughly informed system educators recognize and exploit all intrinsic motivational conditions and states of a learner in their guiding of learning experiences. They do not rely exclusively on detailed planning, continuous observation and direction, or constant reward-punishment manipulation. Of necessity their more open or free instructional patterns require the development of integrated and cooperative multi-loop systems of learning and communications interests, attitudes, content and activity structures. This more open and flexible network of communications and interactions focuses more directly upon the learner and his interests and behaviors than upon the teaching act per se.

Yet the systematic educator will effectively employ the theories and methods of science and system if he and the learner mutually desire to develop units for precise or complex instruction and skill training. In such a circumstance he will carefully plan for built-in instructions, a built-in feedback or self-evaluation system, input overload control, and considerable activity and opportunity to exercise nonfatal and inexpensive choices by the learner. The instructor himself will supplement or reinforce this prepackaged program unit as an available supportive resource. High standards of performance will be expected for appropriately limited periods of such concentrated work on the part of the learner.

In guiding such precise learning experiences, the systematic educator begins with the clarification and improvement of learning objectives. He includes the learner in this task to a maximum degree. Once the learning goals are established and agreed upon, the instructor selectively controls the developing process and the content and experience load. He establishes sequences where necessary, attempting to keep success in task achievement appropriately high. Model answers and/ or alternatives are provided where needed. The pace is set primarily by the student. The instructor manages the learning environment to the degree that unproductive distraction is reasonably and efficiently controlled. And the learner and instructor cooperatively work toward integrating new content into the learner's behavioral repertoire.

In order to illustrate the difficulty of instructional communication, whether it involves verbal messages or demonstration on the part of the teacher, a descriptive paradigm is presented in Fig. 15. It describes a message and two perceptual overlays which contrast the probable differences in the experiential code employed in its interpretation, indicating the difficulty of precise and complete instructional communication.

The efficiency of the translation of a teaching message including its effect can be partially controlled by careful coding (controlling the mes-

The Message as Perceived by the Teacher Is Characterized by:	The Actual Message Is Characterized by:	The Message as Perceived by the Student Is Characterized by:
1. Preidentification of purpose and process, structure and form.	1. Purpose, directional flow, predistribution of code and/or unique qualities:	1. A need for awareness or identification from among many environmental stimuli.
2. Its order of values which is preestablished and credible	2. Form, structure beginning, serial order, and end.	2. An ambiguity of form, structure, value-use, meaning, and credibility.
3. Predetermined channels of distribution.	3. Breadth—the amount of signals presented simultaneously.	3. A receiver who may need help in decoding the message, reintegrating or assimilating it into his program code.
4. A sender who is probably partially aware of its probable outcomes or effects.	4. Length—the number of signals presented serially or redundantly.	4. A receiver who may need help in determining appropriate channels of response or application.
5. Partial preprocessing through a relatively sophisticated program repertoire or code.	5. A coincidence or time dimension. It may require immediate reception or provide for random access or redundance.	5. A receiver who has little knowledge of its value or effect.
6. A synthesis which has cognitive meaning and probably affective meaning (manifest and latent content).	6. Inner and external context, manifest and latent meaning.	6. A receiver who may need help in determining both manifest and latent meaning.
	7. A degree of congruence with a previously distributed code, determining ease of translation.	7. A receiver who may need assistance in controlling the message size, rate of reception, complexity, and sequence.

FIG. 15. A message transmitted by a teacher to a student.

sage structure) and careful transmission (controlling the message transmission). The instructional message permits and usually requires these controls:

1. Controls on the message structure:
   a. The message structure and form (vocabulary and/or experience) will be relatively familiar, easily assimilated, a logical next step.
   b. The message will be limited in size (length, breadth, complexity).

    c. The space-time dimension will be optimally controlled; i.e. messages with great impact or momentum will have controls on length, complexity, order, rate of presentation, etc. Many actions fall in this category.

    d. Lengthy or important messages will be structured for random access, for redundancy by means of repetition or the use of both oral and written form.

    e. Messages involving significant impact will be expedited by controlling pre-conditions, by preparation, pretest, and planning for careful quantity and quality control. Error entropy will be controlled by continuous cuing and support, by relaxing one dimension of quality or quantity control to gain in the other dimension. Planned relaxations of controls may acknowledge an acceptance of trial and error probabilities or the toleration of mistakes. This relaxation clearly stresses the value placed upon the new learning to be acquired.

2. Controls on the message process:

    a. The teacher may provide for a demonstration of the operation to be learned. This in effect displays a system of cues, a model answer. In a complex operation the demonstration may well precede a pupil's active participation. In simple operations, cues, including model answers, may follow the participation.

    b. Distracting stimuli in the environment will be managed.

    c. A participant action-response is expected throughout the process, facilitating continuous feedback. This reduces time loss through errors which are dependent upon evaluation external to the materials at hand.

    d. Although the learning experience is primarily under the control of the learner, the teacher is available as a resource to assist if requested in the interpretation of problem operations involving significant precedences or consequences.

    e. The learning event will be integrated via transference into a variety of situations which will demonstrate its utility and significance.

Instances of the application of interdisciplinary system concepts and techniques to teaching and to the selection and organization of teaching strategies are becoming increasingly common. Among the educators who have written extensively upon the subject and have tested systems theories via research and experimentation is David Ryans, whose pictorial flow diagrams and models of instructional events are perhaps the most comprehensive in existence. Ryans has also conducted important research relating teacher characteristics to learner behaviors.

Law 1. Objectives to be learned are clearly presented.

The Lecture

The lecturer must carefully specify the primary objectives of a lecture or lecture series. He should check carefully for aversive or subjective responses that indicate uncertainty of objectives.

Tests of immediate recall are useful in early lecture patterns.

A pattern of lecturing as a basic method must be supplemented by written materials.

The Group Discussion

A useful discussion presupposes a background of information and basic skill in logical analysis.

A demonstration of trained participants often assists in acquiring such skills.

A teacher needs considerable structure in initial discussions.

Diverse assignments to subgroups can assist in identifying problems and clarifying goals.

Law 2. The content is presented in small information task units.

The Lecture

Restatement of key concepts by pupils using complete sentences is valuable.

Problems requiring application of key concepts help control task size.

The number of task units requires careful control.

The Group Discussion

Discussions can be held to specific topic goals.

The significant property of a discussion is that it generally is not structured in great detail; therefore it cannot be exactly replicated.

Reasonable topic controls, careful time control, summaries, etc. probably should be established.

Law 3. The information task units are arranged in a sequence to maintain high continuity.

The Lecture

If the solution of the first task assists the second, etc., the law is supported.

If the lecture is presented in a series of task units for later recitation or application, error is reduced.

The Group Discussion

Demonstration is often valuable here also.

Points relevant to continuity can be identified by the teacher.

Reinforcement of examples of continuity or relevance is possible.

Occasional summary of progress toward goals is useful.

Law 4. Each task unit contains ample cuing to predispose success.

The Lecture

Demonstration, illustration, verbal cuing of task efforts support the lecture.

Knowledge that material content will assist in the solution of tasks and will be tested later will assist the process.

Important tasks should be liberally cued.

The Group Discussion

Verbal feedback is continuous, instantaneous.

Participants should not be involved for sustained periods without some response.

Manifest as well as latent feedback is possible in verbal discussion patterns.

FIG. 16. Comparative strengths and limitations of two instructional methods.

Law 5. A model answer is available to the student after he has completed each task.

The Lecture

The speaker pauses occasionally for the students to record notes, ask for feedback information.

Applicable printed material, prepared in advance, supports cuing.

Feedback sheets (questions) on the lecture texts assist the student.

The Group Discussion

Recording of discussions permits comparison, reevaluations, etc.

Length of discussions can be gradually increased, with the progress noted.

Small group discussions can report in a pattern of discussion summary by the class as a whole.

Law 6. The rate of pacing is set by the learner.

The Lecture

This is a serious problem in the lecture.

Pausing every five minutes for questions may help.

The lecturer may begin slowly and accelerate.

Students may be asked to submit questions in advance in a lecture series properly outlined.

Reviews of past lectures, main concepts are often productive.

The Group Discussion

The rate is controlled to a great extent by the participant.

An effective teacher assists in adjusting rate control to individual student needs.

Topic or content control is quite easy as one item is usually discussed from numerous perspectives.

Law 7. The process is set up and managed to minimize distracting stimuli.

The Lecture

Extraneous distractions can be controlled.

The lecturer can purposely vary his speed, loudness, tone quality.

Easy tasks can be interspersed among more difficult ones.

Important points can be stressed in voice quality and otherwise.

Some tasks not immediately solvable add to the learning.

The Group Discussion

Extraneous distractions can be controlled.

Some structure toward the recognized group goals usually should be maintained.

Irrelevent discussion should not be permitted to consume too much time.

Law 8. A review test is presented that makes clear to the students the specific behavior needed to meet the objectives.

The Lecture

A preknowledge of test content and eventual test of learning assists this process.

Frequent summary of material aids learning.

In a lecture series occasional review of important points presented earlier is productive.

The Group Discussion

Discussions can act as practices or trials themselves.

The values of constructive controversy should be identified and included in discussion goals.

Opinions, a very real part of discussions, can be separated from proved facts.

The structures of debate, Roberts' rules, etc. can be taught and used in evaluation.

Finley Carpenter and Eugene Haddan employ logical and systematic procedures to illustrate the strategy of choosing among alternate methods of instruction in order to make the choice more functional. They indicate that methodological choices are determined by many input constraints and restrictions. It is highly probable that, at different stages of a complex learning sequence or program, deliberate alternation or reversal of learning activities and mental perspectives is necessary to maximize learning effectiveness.

As an example of the constraints which are built into the common lecture and group discussion methods of instruction, Carpenter and Haddan relate a standard set of learning laws or procedures to operationally tested rules of thumb believed to explain their positive and negative properties. A descriptive model developed from the ideas of these authors (Carpenter and Haddan, pp. 145–172) is illustrated in Fig. 16. It depicts the contrasting characteristics of the two methods in the general order followed in operationally tested and confirmed learning sequences.

A careful analysis of this descriptive model will indicate to the reader the constraining effect that purpose imposes on any operational method or system. Perhaps the presentation of this paradigm will demonstrate further the practicality and utility of system and information concepts and models and will serve to indicate the reasons that educators are increasingly looking to the disciplines of semantics, logic, and psycholinguistics for guidance in the proper or effective use of symbolic codes in interpersonal communications and instruction today. Linguistic theory is a source of great potential for advancing learning and instructional theory.

It would seem that the particular strengths of the two contrasted instructional methods are:

The lecture excels in:

1. Orientation to new concepts, ideas.
2. Integrating facts into larger units.
3. Emphasizing perspectives.
4. Simulating feelings.
5. Demonstrating an orderly, systematic, rationalistic process.

The group discussion excels in:

1. Recognition of issues developed in the verbal exchange.
2. Clarification of problems or goal perceptions (planning).
3. Improving of verbal repertoires, verbal skills.
4. Tolerance for diversity.

5. Evaluation of competing positions.
6. Assumption of various roles.

The works of J. P. Guilford and his associates in the area of personality and the contributions of Benjamin Bloom, David Krathwohl, Bertram Masia, and others in the development of a taxonomic system of educational objectives are significant milestones in the development of a systematic (analytic yet integrated) perspective of the complex educational process. They indicate the extreme complexity of learning and instruction but, at the same time, advance the systematic classification and ordering of some of the elements involved.

Guilford's contributions to education have more or less assured educators that the traditionally oversimplified and overstabilized conceptualization of intelligence and the processes and instruments of its measurement and interpretation are generally to be considered as tentative, inadequate, and incomplete. He and his associates have identified more than fifty-five semi-independent factors in intelligence that can be tested and measured. Among identified factors are those operational factors called cognition, memory, convergent production, divergent production, and evaluation. These operational factors interact with structural or product factors which Guilford categorizes as logical or cognitive units, classes, relations, systems, transformations, and implications. The third dimension of the Guilford model of the intellect contains content or media categories which he defines as figural, symbolic, semantic, and behavioral.

Benjamin Bloom, Krathwohl, and their associates have directed their efforts toward the development of a system of classification or taxonomy of behavioral elements that normally comprise the objectives of a comprehensive educational program. To date their work has produced two volumes entitled the *Taxonomy of Educational Objectives*. Handbook I is subtitled *The Cognitive Domain*, and Handbook II *The Affective Domain*. A third volume devoted to the classification of psychomotor objectives has been proposed also. In their taxonomic developments these authors adapted an organismic approach to education and social process. They seemed to recognize that education encompasses the translation of all human experiences that are important and meaningful, that it involves "knowledge of" as well as "knowledge about." They are aware that education involves valuative, incitive, and formative communication and evocative direct experience, as well as designative and informative communication and indirect experience.

In their work on the taxonomies, Bloom and associates carefully constructed their taxonomic classification systems and logically tested them in order to assure that all of the taxonomic categories and levels were as exhaustive of the universe level and as mutually exclusive of one another as possible. Upper levels in the taxonomic schema were assumed to be systematically more complex. The authors reasoned that lower levels were generally but not entirely prerequisite in the order of their learning.

It is interesting to note that the reflections of Bloom and his associates indicate their belief that the cognitive domain emphasizes rather well-defined, closed, and tested educational objectives—those involving remembering, reproducing something previously learned and derivable by means of reconstructive or deductive-analytic thought. (Krathwohl, pp. 6–7) The volume on the affective domain emphasizes that the more subjective or open learning objectives of feeling tone, emotion, and degree of acceptance or rejection; the affective qualities of adjustment, value, attitude, appreciation, and interest; and the systematic or persistent affective qualities of individual human character and conscience are not nearly so well defined or deliberately induced and attained in an instructional program. Nor are they as easily and objectively evaluated as are the cognitive objectives.

It is significant that the authors of the taxonomies state that the cognitive and affective domains are semi-independent but not exclusively so. They suggest that in general one domain is neither equivalent to the other nor entirely independent of it. They are intertwined, interdependent, and equifinal. They note that the lower levels of learning in one domain tend to parallel and accompany similar levels of learning in the other domain, and they suggest that every emotion may have a cognitive counterpart and vice versa. (Krathwohl, pp. 45–62) In general they support the position that the affective domain is more concerned with self-motivation or *does do* than cognition, which is more closely concerned with *can do*.

In respect to motivation, the authors of the taxonomies note that the higher levels of objectives in both the cognitive and affective domains require more independent and persistent activity on the part of the learner, more self-initiation and self-control of his own behavior. This position is consistent with the theoretical propositions of cybernetic psychologists and the beliefs of those educators supporting greater use of inductive processes and individualized inquiry strategies for optimum learning.

A most interesting analogy is advanced by the authors of the affective taxonomy (Krathwohl, p. 60):

> . . . Thus a cognitive skill is built and then used in rewarding situations so that affective interest in the task is built up to permit the next cognitive task to be achieved and so on. Perhaps it is analogous to a man scaling a wall using two step ladders side by side, each with

rungs too wide apart to be conveniently reached in a single step. One ladder represents the cognitive behaviors and objectives, the other, the affective. The ladders are so constructed that the rungs of one ladder fall between the rungs of the other.

In their development of the taxonomies of educational objectives, Bloom and Krathwohl state that they believe that the overall ranges of the cognitive and affective domains are similar. It would seem probable that, particularly at the lowest and highest levels of the taxonomy, the lower level of primitive and uncertain initial perception and the higher level of internalized judgment, it is difficult to distinguish or separate cognitive and affective behaviors. Systematic decision theory supports the synthesis of cognitive and affective or perceptive behaviors at the intuitive and reflective decision-making levels.

Figure 17 summarizes and compares the taxonomies of educational objectives, cognitive and affective domains, as presented below. The original separate taxonomies are modified in this summary presentation by placing the classifying symbols for lowest or Initial Perception level and the highest or Evaluation level so that they are shown to be common to both domains. In addition the two levels of affective judgments are arbitrarily added by this author, justified on the basis of their consistency with decision theory.

Relating the taxonomies of educational objectives to the task of curriculum development and implementation is another task benefiting from systematic analysis techniques and theories. It would seem that a systematic curriculum plan would provide for the selective alternation of attention to objectives in the two domains and for frequent overlap, redundance, and review of content representative of particular objectives. Such a mixed strategy would seem more likely to achieve the desired goals than one requiring or assuming an exact step-by-step order of alternate presentations or simultaneous ones. Certainly the restriction of educational programs to one educational domain or to the achievement of a limited number of lower level objectives cannot be considered as adequately comprehensive or penetrating.

It seems probable that future educational programs will be developed which are designed to attain higher educational objectives in both domains, but this will not be accomplished within the traditional didactic-deductive educational methodology predominant in some American schools today. An emphasis upon direct involvement and experience, upon self-motivation and self-direction, and upon higher and more complex cognitive and affective processes and learning responses will likely occur.

In the teaching and achievement of affective learning objectives, it is

I. Cognitive Domain   II. Affective Domain

1.00 Initial Perception

I. Cognitive Domain	II. Affective Domain
1.00 Knowledge	1.0 Receiving (attending)
1.10 Knowledge of Specifics	1.1 Awareness
1.11 Knowledge of Terminology	
1.12 Knowledge of Specific Facts	
1.20 Knowledge of Ways and Means of Dealing with Specifics	1.2 Willingness to Receive
1.21 Knowledge of Conventions	
1.22 Knowledge of Trends and Sequences	
1.23 Knowledge of Classifications and Categories	
1.24 Knowledge of Criteria	
1.25 Knowledge of Methodology	
1.30 Knowledge of Universals and Abstractions in a Field	1.3 Controlled or Selected Attention
1.31 Knowledge of Principles and Generalizations	
1.32 Knowledge of Theories and Structures	
2.00 Comprehension	2.0 Responding
2.10 Translation	2.1 Acquiescence in Responding
2.20 Interpretation	2.2 Willingness to Respond
2.30 Extrapolation	2.3 Satisfaction in Response
3.00 Application	3.0 Valuing
	3.1 Acceptance of a Value
	3.2 Preference for a Value
	3.3 Commitment (conviction)
4.00 Analysis	4.0 Organization
4.10 Analysis of Elements	4.1 Conceptualization of a Value
4.20 Analysis of Relationships	4.2 Organization of a Value System
4.30 Analysis of Organizational Principles	
5.0 Synthesis	5.0 Characterization by a Value or Value Complex
5.10 Production of a Unique Communication	5.1 Generalized Set
5.20 Production of a Plan or Proposed Set of Operations	5.2 Characterization
5.30 Derivation of a Set of Abstract Relations	

6.00 Evaluation

I. Cognitive Domain	II. Affective Domain
6.10 Judgment in Terms of Internal Cognitive Evidence	6.1 Judgment in Terms of Internalized Affective Evidence
6.20 Judgment in Terms of External Cognitive Criteria	6.2 Judgment in Terms of External Affective Criteria

FIG. 17. Modification and synthesis of two taxonomies of educational objectives (derived from Krathwohl, et al., pp. 176–193).

important to recognize that problems of professional ethics are involved here just as they are involved in the teaching of cognitive goals. An ethical system educator will not discount or avoid the affective domain in his instructional procedures. Neither will he deliberately create undesirable or inappropriate social dependencies because of his skill in this area of instruction. In guiding learning in the affective domain, the professional teacher will provide varied experiences which are appropriate to the learning desired and the maturational needs of the learner. He will also provide for both teacher and learner evaluation and feedback immediately following each significant affective experience.

When the system educator is concerned with the development of higher cognitive and affective learning processes, he will tend to guide the learner successively less closely. He will alternate more frequently between inductive and deductive teaching, stressing freedom of non-focal or heuristic induction and self-evaluation on the part of the learner. As a learner appears to increase his capabilities in the two learning domains, he is guided toward the higher educational goals of reasoning and valuing. He is encouraged to learn how to learn, to seek and test situations and relations involving positive and negative transfer, and to make judgments wisely.

Education directed toward the attainment of higher cognitive and affective objectives is an art or craft, quite distinct from completely systematized and mechanized processes of lower skill development. Complex learning objectives are taught by considerable variation of mediating activities and responses within broad experience limits roughly predetermined only by the accumulated limits of prior human knowledge. Diverse social and environmental contexts must be provided in achieving more complex learning objectives. As a rule, the efficiency of the higher learning processes will be reflected in the positive rate of self-reinforcement of desired learning responses, reflecting a synergistic and serendipitous motivational effect upon the learner.

An optimally productive curriculum will not be narrowly scientific-cognitive. It will carefully provide for and integrate cognitive, psycho-motor, and affective educational objectives into its total program. Of necessity such a comprehensive and integrative curriculum design will have to be quite selective in the establishing and ordering of goals and experiences in all of the disciplines. Hilda Taba and other curriculum experts have long advocated the careful and relevant selection of learning content and experiences. There is little doubt that it will become increasingly apparent to large numbers of professional educators and their clients that they cannot tolerate the customary economic and psychological waste and political error of teaching irrelevant isolated facts and useless behavioral routines in a meaningless coverage of unselected content much longer.

The inadequacy of traditional deductive, didactic, and doctrinaire instructional processes also is widely recognized by educators. An appropriate use of the dynamic methods of inductive or discovery teaching involving theory-based or structured concepts and active learner reponses is perhaps the most critical educational change which has occurred during the century. We are finally guiding the positive development of thinking and inquiry in some of our more able youth. In the perception of the traditionalists, the educational results appear to be revolutionary. John Dewey foresaw the desirability and probability of developing inquiry methods in education many years ago.

General system input components	Academic disciplines	Applied academic and practical disciplines	Interdisciplinary individual-social strategies for application	Universalizing interdisciplinary strategy	General system output components
Systems Analysis	Humanities				Plans, Records, Reports of
Define output objectives	Philosophy	Applied Philosophy	Imagination		Content
Enumerate input components	Religion	Applied Religion			
Analyze input components	History	Applied History, Libraries, Museums			
	The Arts	Applied Arts, Journalism			
Select input components	Psychology	Applied Psychology, Propaganda, Advertising	Communication		Process
	Social Disciplines				
Select a plan	Anthropology	General culture, Communications			
	Social Psychology (Family)	Homemaking, Practical Arts			
System Design	Sociology	Welfare	Heuristic Trial	Education	Organization
Design prototype model	Politics	Government			
Test prototype model	Law	Law			
Develop, distribute all new units and systems	Economics	Industry, Commerce Transportation	Engineering		
	Natural Sciences				
	Biology	Health, Medicine, Athletics			
Implement fully	Chemistry	Chemical Engineering			
Evaluate	Physics	Mechanical Engineering			
	Electronics	Electronic Engineering			Measurement and Evaluation
	Mathematics	Communications Engineering			

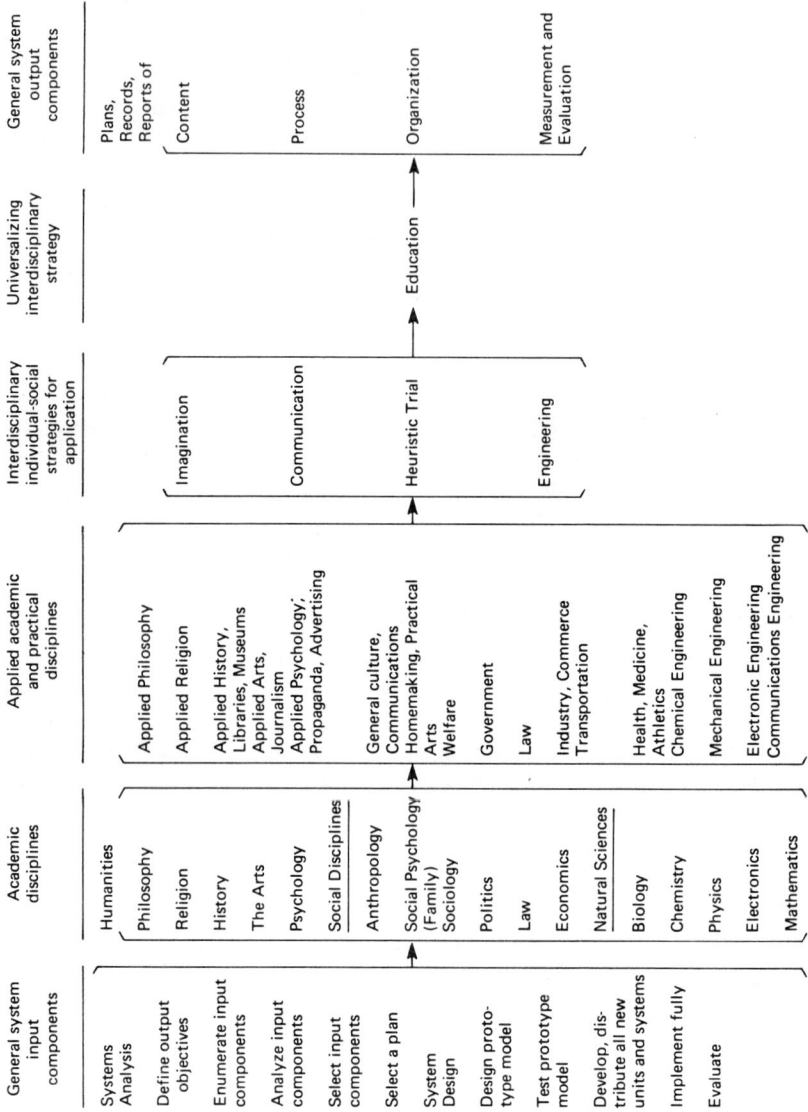

Fig. 18. Universal human institutions.

In recent years, however, a general dissemination and refinement of the methods of scientific and systematic inquiry to enhance its interdisciplinary values have made the process simultaneously more complicated and more precise and economical. These modifications have made the process more applicable to ill-defined and significant problems of all kinds and appear to be guiding education toward a greater learning synthesis and integration once again.

Informed educators today generally believe that a formal educational curriculum in any country which desires continuing world influence, progress, and status needs to be truly comprehensive, not compromisingly so. Historians, anthropologists, and systematic philosophers and educators have pretty well affirmed this necessary relationship. Learning content and experiences must be selected from all disciplines if the curriculum is to be dynamically balanced and integrated. Economies must come from selection within disciplines not among disciplines as far as society is concerned.

Future curriculums must include the presentation of major concepts and experiences representative of the significant structure in all academic disciplines, all psychological needs, and all inquiry and research methodologies. A truly systematic curriculum will provide a planned balance and variety of formative and informative experiences in the humanities, social disciplines, and in the physical sciences. Psychomotor and practical skills will also be adequately and continuously exercised. Advanced levels of affective and cognitive learning will be achieved.

The next figure, (Fig. 18), presents a flow diagram depicting elements to be considered in the designing of a systematic and organismic educational curriculum. It relates the academic disciplines, practical disciplines, and the universal strategies of inquiry and application. The universal interdisciplinary strategy of education is indicated to be the single integrating or unifying curriculum system component.

19

Teaching and Communicating as Art and History

It has been said that science and system always provide a bland, conservative, and short-sighted solution to human problems even if it is a reliable and economic solution. The implication of this statement is that scientific inquiry and systematic technology are not entirely successful in transforming reality in a manner which is entirely beneficial and satisfying to man.

At this time a brief statement is advanced in support of the truly comprehensive curriculum, one which meets all educational needs and allows fully for educational adequacy and perfectibility. Consideration of the input content and experiences and output values of art and history as they function in teaching and communicating will be discussed in contrast to the particular benefits and costs of science and technology.

Education and communications may be defined as the sum of all human experiences and a posteriori wisdom which can be transmitted, shared, and exchanged between an instructional message source and a learner or receiver. It is the ebb and flow of shared human feelings, images, perceptions, and ideas, the pulsation of social momentum. In a comprehensive formal educational program we are concerned with identifying all possible and useful types of human experiences and communications and in providing instruction which facilitates their skilled and appropriate application. A truly enriched instructional curriculum includes the methods of art and history, sometimes called the art of the spectacular or significant, and the methods of science, the art of the informative or consistent.

Before describing these two contrasting modes of general human inquiry and activity, it seems appropriate to discuss briefly those general human capabilities and environmental circumstances which exist before and after the fact of a learning experience, but which are modified by meaningful instructional events.

History and science both tell us that we must acknowledge that existence and creation are simultaneously independent of and dependent upon man. Nature creates and transforms independent of man, but man, unique among all known creatures, has developed a capacity for modifying nature and creating artifacts and synthetic realities (including ideas) which are relatively independent of nature.

Next we must acknowledge those peculiarly human capabilities which assist men in exploiting their efforts through interdisciplinary inquiry and application to determine (1) what is significant in human life and nature, and (2) what is consistent in human life and nature. Relatively independent of the application of a particular strategy of inquiry, humans have the following capacities:

1. Imagination—the power to form mental images of what is not actually present or presently existing.
2. Creativity—the power to construct artifacts or elements which never have been experienced and which do not exist in nature.
3. Intuition—direct or immediate knowing, sensing, apprehension independent of conscious or mediated reflection.
4. Discovery—the first human sensing or revealing of an existence.
5. Memory—the power to reformulate or reconstruct mental images.
6. Belief—conviction or fixing of significant or consistent truths (based on blind acceptance or faith and/or upon observation and reasoning).
7. Wisdom—the ability to learn apperceptively, to interpret from past experience, whether or not that experience was acquired perceptively or reflectively.

It seems likely that the most natural or primitive of the above capabilities of man are intuition and discovery, followed perhaps by imagination, creativity, belief, memory, and wisdom. However, there is considerable doubt about the degree of advancement of these general human capabilities prior to the development of the greatest human tool or instrument of all, language. They are probably all interdependent upon human linguistic and communications abilities.

Through the invention and perfection of his language systems, man has refined and extended his prior mental and physical abilities, including his ability to perceive, sense, and believe. Language was and is an instrument for perfecting memory, belief, and wisdom. With the development and refinement of language man has advanced his ability to generate and use:

1. Ideas—mental images or conceptualizations of reality.
2. Intelligence—the ability to learn from experience moderated by reflection and reasoning.
3. Idealization—the creation of archetype ideas, visionary or perfect models.

The first significant advancements of primitive societies into a higher-level or more complex society followed shortly after the development of written symbols to represent spoken language. During the classical periods of history men employed language together with observation to refine their communication and their methods and rules for representing reality and its natural objects and relations. The products of the first civilized applications of human artistic, historical, and reflective method were *artifacts*—objects transformed by human effort, and *knowledge and wisdom*—the recorded language and history of human inquiry and activity.

The twin human strategies of applied art and deductive science continued to be the dominant strategies of human inquiry and application down to the seventeenth century or so. The Renaissance was that period in human history which marked the culmination of this age. During this entire period changes in the rate of accumulation of knowledge and in human inquiry strategies were slow.

When the method of human inquiry was extended to include the experimental method, the method of imperfect but controlled induction, the activities of men and their relationships to nature began to change more rapidly. Science has accelerated inquiry, refined all of the language systems, and complicated all of the disciplines. The natural and biological sciences and mathematics in particular have become sophisticated and complex. Among the social disciplines, economics and psychology have become near-sciences. The social disciplines of politics, sociology, and anthropology have emerged as important new entities. History and the arts have assumed a lesser and more specialized role in the mainstream of human affairs.

When men study and compare the history of the methods of ex post facto art and history versus experimental science, they cannot help but hypothesize that perhaps art and history are decadent strategies of human inquiry and activity. Perhaps these "natural" strategies of human inquiry are applicable only to primitive societies, to eras when recorded human knowledge was relatively incomplete.

But perhaps this is too simple a characterization. Perhaps the strategies and products of art and history are useful in explaining and modifying those systems and problems which are as uncertain and complex today as they were in the past. Perhaps the methods of art and history explain and model dynamic social systems more truly today in some respects than the more complicated but limited systems of science and technology.

Art and history continue to be highly reliable codes for representing and interpreting complex and dynamic social systems, systems which are highly unpredictable and open. They continue to record faithfully the significant parameters of possibility and probability in human-ecological affairs. They are often the best available methods for determining the primary values of human actions—the who, what, to whom, and why parameters or properties of events.

Science was and is most effective when it is employed as a model of relatively closed or determinate systems and operations. It is useful in explaining and controlling events and systems in which all components and operations are relatively perfectly enumerated and whose behavior and relationships can be assumed to be consistent and well-ordered. Science as a model of extremely complex dynamic systems errs in its assumptions of the completeness of knowledge, the degree of consistency in the system, the assumed precision of prediction, and the fact of assured control. The more complex and dynamic the system, the less faith we can assign to the probable utility of scientific and mathematical models as consistent predictors and simulators, let alone controllers.

If men compare the input processes and the short-term and long-term output products of art or history and science, they will find that they tend to contrast consistently in these characteristic ways:

Art and History	*Science*
Art and history are based in human perception and produce wisdom.	Science is based in conceptualization and produces knowledge.
Art and history tend to model the organismic, the macrosystem, the long range.	Science tends to model the particular, the atomistic system, the short range.
Art and history invent and record artifacts, wise or significant events.	Science invents and records knowledge and facts; it reconstructs artifacts.
The long-range or systematic function of art and history is to create and determine the significant in creation, perception.	The long-range function of science is to invent theory and determine the consistent in creation, perception.
Art and history demonstrate a balance of human skills, the continued significance of human perceptual, psychomotor, and cognitive skills.	Science demonstrates a specialized reliance upon cognitive skills; it tends to employ technology to replace or extend human psychomotor and perceptual skills.
Art and history epitomize the unique.	Science epitomizes the reconstructed, standardized, replicated.
Art and history generate new tasks for men.	Science frees men from undesirable, boring, dangerous, or painful tasks.
Art and history maintain human significance.	Science tends to advance technological significance, creating a dependency among men.
Art and history conserve and maintain human customs, methods, and means.	Science complicates and enriches choices of methods and means.

AS ARTIFACT

Human Skill and Craft	*Technology*
Stabilize production rate	Accelerates production rate
Permit unique productions, limited distribution	Permits mass production, distribution, standardization
Create new and precious artifacts	Creates economical artifacts, with low cost per standard unit
Emphasize the intrinsic, the impractical	Stresses utilitarianism, practicality

If educators systematically compare the above input costs and output benefits of the methods of art or history and of science, they will perhaps better understand why science is considered the bland but sure strategy of human inquiry and application while art is the liberating but high risk strategy. However, if the full ebb and flow of human developments and progress is to continue, it appears that the two basic human inquiry strategies will continue to be necessary and complementary in some degree of appropriate balance. Any educational curriculum relying exclusively on only one of the above major inquiry strategies will prove to be debilitating in the long run.

Now let us consider the disciplinary content and response strategies of art and history as media with considerable utility in current social communication and human affairs. In its varied forms, human communication may be transmitted through action and language directly and indirectly, consciously and unconsciously, intensionally and unintensionally. In education, even when using linguistic communication, we are concerned with both intensional-expressive-valuative communication and extensional-designative-informative communication.

Intensional communication, the initiation and presentation of a communication or experience which is complex, uncertain, and formative, requires learning responses which must include the perceptual and psychomotor responses—action, involvement, and participation. According to authorities such as Bloom, all of the higher levels of cognitive learning require such experiences and responses also. Inductive or discovery teaching is a deliberate representation of intensional communication. Intensional or evocative communication, direct action, and inductive teaching can only expect heuristic or search responses, reactive expression, relatively uninhibited, evaluative-free, risky, and uncertain responses at first. Only after considerable search and trial does inductive learning produce adaptive, purposeful, and efficient responses.

Inductive teaching in science is intensionally modified by means of systematic rule building and application to achieve the controlled induction or experimental pattern characteristic of the scientific method. Any social discipline which uses rules of organization to systematically and incrementally modify responses to inductive situations and intensional communications is to that extent a science.

Purpose-forming inductive experiences and intensional communications are primitive art. As art they express the whole truth of significance and feeling and require in their comprehension and response intuition and perception. An art experience is a unitary synthesis of problem and solution fully infused with values. The experiencing of art gives the receiver a sense or taste of being, for a moment, universal or cosmic man. He is in full and complete communication with the situation.

Only through initiation and response of intensional communication does a human being sense and fully perceive the apprehension of originals or singulars. Only through direct experience and intensional communications does he have the opportunity to have an intuitive extension of consciousness, an image or paradigm of a new reality. Through art and history men exercise and record their intransitive experiences, the creation of their archetypes or ideals, the foretastes of their discoveries and inventions. (Read, pp. 44–158) Intensional communication as displayed in art and history reveals as its primary necessity itself. It is subject and object, a complete product. Its values are intrinsic and independent, not extrinsic and utilitarian.

Extensional communications (deductive, didactic, and directive instruction) are dependent upon both a past and a future state, condition, and action. Science and informative-cognitive communications are primarily process-centered. They initiate and generate responses which are laden with extrinsic and utilitarian values only. It is no accident that logic and mathematics are particularly applicable to the methods of science and the particular social discipline of economics. They are all devices of extensional communications which produce only utilitarian values.

The informational significance of an art form or intensional experience always follows its creation at a relatively slow pace. Art and history are slow media. The experiencing of intensional communications moves the learner only slightly forward into the realm of meaning, including the domain of exact symbolic representation. However, inductive and intensional experiences are eventually and cumulatively susceptible to more precise interpretation and meaning, as the rules for their description, explanation, prediction, and control are invented or discovered.

Where inductive and creative experiences generate significance in the form of universal symbols or universalizing experiences, deductive reconstruction and invention generate precisely consistent universal symbols for representing these identities. Art creates significant truths; science recreates consistent truths. Both types of learning inputs and responses are necessary for a continuously developing dynamic and progressive society.

Historically, artists, idealists, humanists, and certain humanistic philosophers and historians have concentrated their interest and inquiry in the disciplines which represent the formative, original, personal, passionate, organismic, and transcendent in human behavior. It is therefore not surprising that most of the great systematic philosophers, historians, and semanticists have sensed and valued art as an active force and a legitimate educational program or group of programs. Only the intensely short-sighted and impractically practical men do not understand the functions of intensional communications so aptly demonstrated in art and history. Dewey commented upon the interdependent relation between art and science or system in this manner (Dewey, 1960, p. 156):

> The enemies of the esthetic are neither the practical nor the intellectual. They are the humdrum, slackness of loose ends, submission to convention in practice and intellectual procedure. Rigid abstinence, coerced submission, tightness on one side, and dissipation, incoherence, and aimless indulgence on the other, are deviations in opposite directions from the unity of an experience.

Any systematic philosophy of education is incomplete without the exercising and the understanding of the esthetically sensed, intrinsically valued, and intensionally communicated. Relatively open inductive learning is essential to maximum human achievement. Only through the disciplines of the arts and history is this type of learning recorded for eventual retrieval.

In the last two decades a number of psychologists and educators have been probing and progressing in the study of human formative experiences, affective and perceptive behavior, and intensional communications. They have tended to concentrate at first upon the investigation of creative behavior ex post facto, as might be expected. Psychologists interested in the study of human motivation, divergent behavior, and extra-rational or pre-logical systems are among this group.

The literature and research of these investigations generally report the discovery of a condition of systematic openness, of low predictability in relation to an immediate or present event and in relation to prior and subsequent systematic behavior. They describe discovering a system in a state wherein the system is moving inductively and intuitively from the probable possible (the path of prediction, expectation, science, least resistance) to the improbable possible and perhaps even to the probable impossible and improbable impossible.

The study of formative experiences and intensional communications is characterized by a low order of determinateness. It involves the improbable prior distribution of the information code. For this reason often only an elite (those persons who have considerable similar experiences and training) can accommodate the first or initial discoveries of meaning and structure in such experiences and in the disciplines which record them (history and art).

An analysis of the conditions which accompany individual formative or truly inductive experiences offers these general conclusions or principles. The optimum conditions for creation and invention have been found to be a high level of general systematic equilibrium or security in the learner but a discernable level of problem insecurity in the area or direction of the creative effort or inquiry. There is thus some focus to the inquirer's efforts rather than merely a general disequilibrium, feeling of tension, or entropic sensing and intuiting. Strong motivating interest guides the learner in the general direction of a belief or goal condition or definition. In this rather unsystematic exploratory action state, effort is maintained and sustained through unsuccessful trial and negatively motivating failures by the original internalized belief or motivational state. History records the major examples of these instances of significant and critical human accomplishments. They are recorded and communicated in the great works of art and the major scientific and social literature.

The transformation of significant individual creations, discoveries, and inventions into formal processes of education appears to progress through four stages which can be described as follows:

1. *Initiation:* The first stage of basic inquiry is the formative stage. It may originate as accident, sport, or as the presentation or recognition of an important and ill-defined problem. The first stage of significant creation or discovery requires the introduction of a condition of intense motivation to sustain search and combat failure. If the search is based upon any prior meaning, belief, or experience (and this is likely) it will have some focus or direction. As an instructional or learning act this stage involves the presentation and acceptance of a challenge and complementary conditions of individual awareness and interest.

2. *Development:* Sustained by primitive conditions of awareness, motivation, and purpose, the learner proceeds to define, invent, act, and evaluate, then redesign and repeat his search strategies and tactics. The evidence from his trials either reinforces his belief and motivational conditions or redirects them. Some degree of sufficient success maintains both the motivation and the direction of action. It is possible that eventually a satisfying goal is attained and the mediating process associated with it adopted as an inquiry strategy or technique.

3. *Diffusion:* Personal or perceptual knowledge and satisfaction acquired by freely exercised inductive behavior feeds upon itself

(grows). A successful or satisfying personal experience will be reconstructed very likely and eventually demonstrated and communicated, either accidentally or purposefully. The experience now becomes public rather than personal. Some of the information code is now distributed.

In formal education the diffusion stage characterizes the process of instruction. It is normally a stage involving innovation, not creation. In deductive-didactic teaching a minimum degree of innovative instruction may be accomplished by establishing doubt and uncertainty in the learner and by employing external motivation in order to interrupt the learner's present equilibrium. This can be supplemented by direct demonstration and by a mixture of evocative and informative communication.

In inductive instruction and learning, the learner is guided more directly by a problem situation which arouses his attention, interest, and response, than by teacher-mediated stimuli. He is given more opportunity for heuristic trial. His responses are not controlled or inhibited so indirectly and continuously by the teacher. He is made more self-reliant and independent.

4. *Adoption:* The goal of education is to change the learner, not only during an event but usually in an irreversible and transitive manner. The behavior experienced in a learning event is optimally retained, reconstructed, and transferred to appropriate situations to be experienced by the learner. Appropriate transfer is assured only if the learner accepts or believes in the value of the behavior, understands and controls it consciously, deliberately, and independently. Such an adoption of innovative learning experiences requires self-identification and personal feedback on the part of the learner, entirely independent of external guidance. Adoption of learning behavior requires that it be proved to be personally satisfying and useful. The learner must be involved directly.

It is much easier for the self-directed and self-motivated learner to be captured by an experience than it is to externally manipulate or control his capture. Systematic psychologists and educators are aware of this advantage. The systematic educator also understands that learning to be retained and transferred must be perceived as significant by the learner relatively independent of his relations to the teacher. This is the primary advantage of inductive or problem-centered instruction.

Yet, as a systematic inductive learning program employs more and more direct or inductive instruction, the learner will become aware of the mutuality of goals and interests he shares with his fellow students and his teacher. He desires to share and communicate his significant experiences. In this way his self-motivated and self-directed inquiry tends to become socially tested and adapted. Eventually his personal and social interests may be perceived as fully compatible.

It is no accident that during the present period of social development, when American technology appears to be the overwhelmingly successful human strategy and when scientific and systematic subjects are dominating educational curriculums, inductive instruction is exploited in the scientific and systematic disciplines as well as in the social disciplines. Educators and social planners and leaders are frequently applying the speculative or freely inductive methods of drama, scenario writing, nonmathematical gaming, and simulation or role playing in the modeling, representation, and planning of social systems. The methods of more controlled induction as reflected in experimentation, mathematical gaming, and operations research are dominating modern programs in complex problem solving in the natural sciences, business, and economics.

The knowledge that men are acquiring about the psychological properties of the fine arts and humanities will probably be applied to formal educational programs in the future to a much greater degree also. The importance of dynamic affective learning as acquired through literature, drama, the fine arts, festivals, sports, and games is becoming better understood in spite of the dominance of science and technology in modern living.

Festivals, games, and recreation are described by some as artifacts, innovations, and popularizations of the sport and drama of life (McLuhan, p. 233). Perhaps this explains why games, festivals, and other forms of recreation or fantasy which serve a nonobjective or nonproductive function in human affairs are particularly capable of refreshing the human personality, releasing expressed reactions against the main drive or system of a culture.

As instruments for instruction, games and drama appear to possess considerable unrecognized educational potential for augmenting symbolic communications. They are useful in initially reconstructing and disseminating certain aspects of affective and emotive experience and meaning. Games can transform the ordinary into sudden luminosity by emphasizing the dramatic, fantastic, and unexpected—the very opposite properties emphasized in science. Thus games and drama may have considerable undeveloped motivational value for the exploitation of inquiring teachers. It would seem that the more open inductive processes of drama and nonmathematical gaming possess the potential for facilitating personal involvement, identification, and free expression to a greater degree than more formal and systematic educational methods.

Games also serve to encourage risk taking, curiosity, and creativity in children, reducing their fears of change, failure, and error. They have inherent in their design and intent an initial simplicity and impact which permits and requires subsequent analysis and resynthesis only after interest, attention, and commitment have been firmly established.

In summary, the argument and rationale advanced above are intended to suggest that the utility of the disciplines of art and history will continue in human affairs, serving as an ever-present complement to the processes of science. Men will continue to need and to have both transformative and informative experiences. An adequately comprehensive curriculum in the American schools needs a full complement or range of disciplinary experiences which exercise all of a learner's perceptual-intuitive, cognitive, and psychomotor skills and abilities. Our society will deteriorate when and if freedom of expression, festival, drama, fantasy, value, and the aspiration to enjoy are replaced by cognition, intellect, technology, by aspirational, social and environmental consistency and uniformity.

20

A System Perspective of an Instructional Organization

In the design of any macroanalytic model of an educational organization the ultimate system model would be immeasurably complicated yet elegantly simple. It would indicate all of the minimum essential elements and operations and exclude, eliminate, or close the system to the remainder.

If one were to analyze a school as an instructional system, he would describe it as a multi-loop dynamic cybernetic system comprised of administrators and teachers as actors. Purposes, materials, and information would constitute other input *operators* in the primary system. The system would include a network of *operations* (action and decision events). And it would include students and perhaps their parents as *operands*, secondary subsystems which the primary system would attempt to purposefully transform by means of predefined operations.

The primary and secondary subsystems in the school are both semi-independent or closed-loop self-regulating human systems functioning in some sort of dynamic interaction. Each possesses unique, independent, and undistributed information in its control system as well as some common code which is that factor which links all of the human actors in a systematic purposing effort. Dual-loop cybernetic networks are the minimum human interface in a meaningful instructional event.

The actual flow of events in an educational organization may be perceived as an operating communications system. A perfectly preprogramed or completely closed educational system would efficiently communicate or translate the input resources, including the educational purposes of the primary system, into the output products of the responding system. The learner would be a cheerful robot, doing exactly as he was told. He would have only limited adaptability and could efficiently translate or communicate for a limited time and in a limited manner.

A more open and dynamic educational system or network would not be perfectly preprogramed. It would successively initiate heuristic or search actions which could and would update or improve both the program goals and implementing operations simultaneously or alternately with the operation of the current standardized program. It would have the ability to improve or regraduate both the cybernetic control and operating systems, perceiving them as equifinal conditions or variables.

The flexibility and capability of a semi-open or quasi-rational social system is infinitely greater than a cybernetic machine system serving the same control functions. This flexibility and adaptability is both a benefit and a cost, depending upon whether the properties being assessed are measured on a scale indicating immediate or standardized accomplishments or on a scale indicating longer-range and progressively cumulating benefits.

The method of system analysis and design or recursive programing is sometimes called value engineering. If we value-engineered an efficient educational or instructional system, we would design the following elements into a symbolic map or model of ordered input elements and operations:

1. Personnel capable of efficient purposing and possessing the technical and human skills for attaining the organizational goals.
2. Special-purpose and general-purpose machines incorporated into the system at any point where they can perform faster, more reliably, and more economically than human hands or minds.
3. Materials and instruments designed to simplify and expedite data gathering and transmission.
4. Discrete items of data (symbols or records representing elements and events) which are identified, defined, and coded.

An effective preprogramed educational system would identify and adopt efficient standardizing rules (policies) and procedures capable of providing the administrative network with:

5. A maximum of essential information (actions and decisions) and elimination of the rest.
6. A smooth flow of information from the source(s) to the point(s) of utilization. Efficient system design ascertains who needs to know what, why, when, where, and how much, and then programs this pattern of information flow and storage.
7. A continuous and efficient progression of operands (students or

clients) as inputs-outputs through the sequence of ordered events in the system and a corresponding distribution of operator resources.

The ideal instructional system would also provide for the maintenance and updating of its own command-control network or program mapping system. This requirement would include:

8. A continuous updating of all relevant information.
9. Immediate access to a memory repertoire of historical records of prior pertinent events and experiences.
10. Documentation sufficient for replication of past operations, together with a means of accommodating desirable modifications into the operating and control systems.

The reader should remember that simplification and economy are desirable and necessary elements in the analysis and design of social organizations or systems. This is especially true when automated data processing produces and distributes information in such abundance. System processes have a capacity for inducing simplicity and economy into educational systems in the following ways:

1. They identify and order the essential objectives, permitting the elimination of the nonessential.
2. They facilitate the controlled induction of alternative methods/ means and provide a basis of selection among alternatives.
3. They permit the weighing or balancing of proposed changes by comparing them with current operations.
4. They expedite the assigning of cost-benefit indices to alternative programs or proposed changes in advance of making expensive and irreversible commitments to their adaptation.
5. They systematize the decision rules as well as establish the command-control communications storage centers (determine the locus of positions and records).
6. They predetermine the design of the evaluative procedures and instruments to be used in monitoring the feedback of operations. This will facilitate almost continuous and current error reduction.
7. They facilitate the study of extremely complex systems of incremental and long-range changes and expedite the forecasting of long chains of decision events.
8. They provide for the optimum integration of all variables including different human perspectives into the inquiry and operations processes.

Without a desire and an effort on the part of the operators of present educational systems to eliminate that which is useless and costly in present systems and the freedom from environmental restrictions regarding this elimination, there is little hope of achieving the full benefits of systematic procedures in educational planning and programing. Such technologically effective, economically efficient, and politically and socially acceptable control of change and innovation is essential for the continuous and cumulative improvement of the educational system.

One further thought regarding the analysis and programing of educational systems is advanced here. If schools have some elements and operations which are more definable, obviously manipulative or causal, and therefore more easily explained, mechanically controlled, standardized, and automated than other subsystems, questions of the appropriate application of technology are pretty well answered. The only remaining planning decisions in this respect are questions of where, when, and how much—all practical questions. Knowledge of the impossibility and impracticality of decisions regarding the selection of technological or methodological means or media thus can be deduced and calculated from accepted system analysis or operations research propositions and need not be left to heuristic experimentation alone. Thus a knowledge of and appropriate application of system analysis procedures will be of practical value to the educator who wishes to predict more precisely the real worth of any technological innovation he is considering for adoption.

THE LIMITS OF PROGRAMATIC MODELS
OF EDUCATION AND INQUIRY

The process of educating children in schools is a problem involving the specification of objectives and a program of facilitating experiences, actualizing these experiences, and then stabilizing the output conditions of both the operator or instructional system and the operand or learner system through positive evaluation and standardization. In natural and mechanical systems the establishment of original patterns of elements and events builds in or assures the consistency of the output state; the outputs of such systems remain generally invariant if the inputs are carefully controlled. In intelligent yet inquiring cybernetic systems, both nature and purpose build in some degree of order, but, as some input variation or purposing is both possible and desirable, variation in output is also possible and desirable. Thus the control of an active learner is never fully independent of his own desires.

For this reason educational programs for influencing human minds and actions must be quasi-controlling, not absolutely controlling. They must depend upon the voluntary contribution of some degree of common attention, interest, active cooperation, and shared evaluation to effect immediate and long-range or stable learning output responses. Perhaps the most important principle that system theory offers the educator is this proposition: human beings must acquire immediate intelligence and subsequent wisdom cooperatively. They learn and communicate only by means of mutually experiencing and monitoring learning events and systems. In fact, it is necessary to mutually establish common preconceptions of goals (beliefs), common observations or experiences, and common feedback signals before the learning intent of an instructor is fully translated into the response behavior of the learner.

In the highly ordered system of traditional society both nature and human behavior were more determinate and/or more random than at present; therefore they were subject to natural evolution or slow change. Educational goals and values were relatively constant under such conditions. Values and value-terms were simple, well-ordered, and relatively absolute. Life and education were simple.

With the progress and refinement of the methods of scientific inquiry and applied technology, systematic change became more central to human experience and to education and instruction. Educational risk became the acceptable pattern of inquiry. There was less fear of change because knowledge of unexpected consequences was built into a method of regulation, and critical accidents were not perceived to be so threatening.

As men are predisposed to risk creating, inventing, and discovering, they are also predisposed to educate their young toward the effective reconstruction of these skills. After some two hundred years of applied scientific method and evaluation of its results, we are beginning actually to teach the learner to exercise the processes of inductive inquiry to some degree. We are finally teaching the student to inquire, to learn how to learn.

However, our present emphasis on inquiry does not mean that in most instances learning experiences are not systematically guided. Systematic learning requires that heuristic trials are based upon a series of successive controlling input-output behaviors. Men's significant problems are ultimately subject to repeated attention and attempted resolution. In systematic inquiry he seeks to determine what conditions in an event or system (the input elements and operations) are system-based or irreversible and which are event-based or reversible. Herbert Simon (in Pigors, pp. 86–101) writes that accurate prediction and control require the identification and assessment of those factors that are the unmoved movers or irreversible event-based constraints including purposive parameters and those which are the equally unmoved invariants—the natural and technological restrictions controlling the system. These are the necessary and sufficient properties which we seek to control in learning experiences. They are the properties which must be reordered in order to establish significant change.

Systematic instruction is a matter of deliberately varying in some pre-specified manner the input elements and activities in a learning environment. Adequate experimental control permits the ascertaining of relationships between instructional inputs and learner responses. In time the lack of knowledge in this doubly mediated communication act is productively reduced. More effective controls can be established. Eventually an event can be more precisely predicted on the basis of assessing the input conditions, and it can be readily reconstructed.

The educational problem of the highest order is to develop a set of communications rules which expedite and facilitate the appropriate exercise of the learner's will when he is subject to his own self-control. This involves the problem of affecting self-motivation, of internalizing values. The critical and ultimate educational mission is to develop and transmit a system of cooperative human behavior responses, a system of values.

It is becoming apparent to both the informed educational system analyst and the casual observer that the precise and determinate control of mechanized production processes is a complex matter in itself. However, comparatively determinate control of individuals and organizations is an infinitely more complicated and uncertain process. Humans are indeed adaptively rational rather than completely rational. Thus it is easy to understand why the simulation of control of human or organizational behavior is extremely difficult.

Primarily because of the great complexity and uncertainty that is an inherent condition of human behavior, educators cannot discount the importance and utility of material artifacts, machines, tools, and instruments in instruction and inquiry. For technology is mutually interactive with and contributory to the development and stabilization of new and improved human theories and processes. Physical instruments for controlling and measuring distance, weight, temperature, rate, and time have been essential to human progress and advancing civilization. Technology is both bridge and transportation system in the human production network which is intended to attain the future in the present. Technology is a veridical and accelerating transformer of human purposes.

And, in addition to physical and material technology or hardware, it is important for the educator to recognize the significance of man's greatest and most common veridical instruments—his symbolic communications systems or software which so effectively simulate the real world. For, when man combined his systems and rules of logic, mathematics, and experimentation with his technology, he devised machines which can perform symbolic transformations or calculations that in turn regulate machines engaged in material transformations. They can also simulate the human capacity to systematically think, calculate, and decide. Without an understanding of this linkage between human symbolic and knowledge systems and human technology we are substantially ignorant of major human production efforts in modern society.

Among the problems of systematic educational analysis and planning is the problem of selecting only the necessary. This is the problem of simplifying and ordering priorities. Logical and systematic analyses are also methods of isolating data conditions which are necessary and sufficient in any purposeful operation. However, logical analysis remains rather complicated when planning combines the application of logical, teleological, technological, economic, and political-social criteria or rules in the making of an intelligent decision. For this reason, and because human attention and memory capacity is quite limited, men must rely on their computer memory and transformation technology for assistance in the resolution of many complex problems today.

Yet men must constantly assist their automata even in routine production, because at any one time the limits of preprogramed knowledge are soon revealed. And in the gradual resolution of ill-defined and extremely complex problems, human imagination is much more essential than information. Only relatively free induction or heuristic search can move inquiry forward in such instances.

In the establishment of any semblance of order or system in such an extremely complex human process as education, the first requisite step or progression requires the determination of output objectives. After objectives have been ascertained, the identification and ordering of a series of input-output transformations or learning sub-goals is necessary. These ordered learning experiences must work, for the limits of function directly affect the accomplishment of the learning program.

Following the effective design of a learning or instructional system, it is essential that it is materialized and actualized. The potential conditions may be present, but the requisite penultimate actualization or final catalytic action may not occur. And it is at this stage that the learner has almost total control of the instructional act. Thus cooperation is necessary to produce the positive self-regulated motivation in the learner if instruction is to be successful. This is the key difference between educating a dynamic cybernetic human being and programing a computer. The assumption that the education of children and adults is a fail-safe or determinate process ignores the human qualities of spontaneous and deliberate behavior.

The limits of variability and cybernetic control (self-control) in instructional systems are also important boundaries to ascertain and maintain. With the development of quantum theory, natural scientists have concluded that even the most basic of natural materials and forces are more inconstant than was formerly believed. When human psychological and communicative variabilities are added to natural and technological variability in systems, the problems of control are infinitely compounded. It is just this recognized uncertainty dimension in the character of social systems—including educational systems—that challenges the interest of the system analyst. For he recognizes that such circumstances almost prescribe the use of techniques of system analysis and design as the only practical method of advancing human knowledge and human welfare beyond their present state.

Finally, the limits of educational feedback systems of evaluation and interpretation also constrain the educational process. Both a priori goal definition and a posteriori act and action evaluation are necessary for incrementally advancing the process of education.

It has been asserted by some theorists that science and system can by-pass or replace human intuition and imagination in inquiry, in the fixing of beliefs, and in productive human accomplishment. This assertion fails to acknowledge the necessary prerequisites and post-requisites in any intelligent system—the unique human capabilities of individual and collective human imagining, purposing, believing, observing, and judging. It does not adequately explain the properties of human persistence and motivation displayed by men as they maintain the difficult and discouraging heuristic operations often necessary for solving significant and ill-defined problems. Nor does this assertion account for the development of widely shared and communicated social, legal, and moral conceptualization or value systems which structure most human acts and decisions.

The best that science and system can do in human inquiry and productive action is to induce workability, precision, speed, consistency, and economy into communications and technological operations. Science and system are and will remain a maieutic (midwifery) or mediating methodology or technique which merely advances, improves, and/or accelerates the reconstruction or transformation of human ideas. This is, of course, no small accomplishment as it is the essence of systematic purpose-oriented action. All deliberate or systematic human acts or technologies are dependent upon the establishment of a priori and ex post facto goals and value determinations.

Scientific inquiry and system analysis procedures can explain but a limited part of the past and predict or control but a limited part of the future. Inquiry or research of any kind is advanced only by the imperfect induction of human ideas. It is merely disciplined and regulated by science and system. (Mills, p. 71) The utilitarian scientist or technologist generally avoids the problems of assessing the latent or indeterminate meanings of events in systems, the less obvious and more abstract action values rather than the immediately determined act values. He concentrates on skill and method at the expense of a more comprehensive value ordering. The implications of many human accomplishments are well outside the domain of such immediate scientific and systematic means and concerns.

The scientist-philosopher takes a broad view in his inquiry process, however. He fully recognizes the tremendous power, judgment, and responsibility attendant to significant systematic inquiry and instruction. He is aware that systematic inquiry has progressed to the point where men can and do create symbolic models for yet to be created material entities of great cruciality and complexity. And he knows that, with increased control of natural and technological means, the hypothesized or imagined end goals of humanly controlled behavior become ever more important.

The broad perspectives of Mills, Dewey, and other systematic philosophers are not always exhibited in present-day efforts toward applying systematic theories, psycholinguistic methods, and information technologies in the improvement of formal education programs for human beings. Many educational applications fail to acknowledge or account for the dynamic openness of human behavior. System models or disciplinary analogues befitting neatly closed technological systems or theories are far from being truly representative of actual human thinking and behavior.

Educators and system technologists should remember that computers are consistently determinate and preprogramed rather than dynamically and indeterminately programed. They are substantially intellectual morons and cheerful robots, and they operate wholly outside the affective and imaginative domains and the domain of intrinsic valuing which characterize human behavior. All of these conditions suggest that the design and management of computer and automation systems is considerably different from the management and education of people.

As controls for automated mass-production, for vehicle guidance, and for the simulation of well-defined events and systems, the computer, with its miraculous high-speed calculative ability and infallible memory, is of the greatest significance. But systematic technology has realistic operational limits of applicability within any one event and within any identifiable environmental context.

In order to maximize the value of technological processes and system procedures, system analysts and educational planners need to work within a fully eclectic or comprehensive programing perspective. This frame of reference must account for and accommodate both human and natural conditions and qualities. Thus no responsible cybernetic theorist has really conceived of a comprehensive educational program for human beings in which their behavior, including their creativity and imagination, their affective-valuative behavior, and their incitive and valuing communication is reduced to purely designative and prescriptive process and action.

In summary, the responsible educational system analyst and design programer recognizes and acknowledges his professional limitations and responsibilities. He admits the existence of limits of both information and technological capability in any conceivable method or system of instruction. He knows that the success or the failure of instruction is shared by the instructor, the learner, and the environmental systems restricting what is taught and how it is taught. He realizes that a teacher at best can only effectively present knowledge and demonstrate his personal and positive belief and application regarding it. And he knows that, as the learner progresses in his education, he gains an ever increasing capacity to exercise his own free will in the acceptance or rejection of information and in the making of action-guiding decisions which ultimately determine the persistence of present personal and social perceptions.

COMPUTER AND MACHINE VERSUS HUMAN CAPABILITIES

The history of human affairs is a record of man's efforts to simplify, organize, and stabilize his physical and mental environment, to carefully transform one or the other of these environments, and then to correlate or regraduate the two environments. Such homeostatic control development is evidenced in the writings and artifacts of civilization.

At the dawn of civilization men ceased their nomadic ways, thereby establishing the possibility of more order in their living. The development of written language and skilled craftsmanship provided an additional basis for organizing, specializing, and standardizing human mental and manipulative skills. Artisanship and eventually machine manufacture continued the stabilizing and standardizing process, for technology is a process which precisely reconstructs events and operations. The industrial revolution is evidence of the evolution of human systematization of control over elements of the environment. Mass production, standardization, and economic efficiency are derivatives of the systematization process.

The evolution of environmental control has been and will continue to be limited by or dependent upon human desire and purpose, as well as by the ability to do. Economic factors are additional regulators of social and technological change. Men, in general, tend to induce change into their social and technological processes only when they can foresee or predict rather obvious or near-term benefits or profits. They exhibit a constant effort to make simplifying and utilitarian decisions rather than truly optimizing decisions. And the induction of change is not without its psychological as well as its logical constraints.

If we consider the evolution of human inquiry we likewise see in its history a pattern of imposed simplification, stabilization, and habituation. Myth, religion, and law are belief systems which attempt to explain and control behavior in simple and stable ways. Ethical, moral, and social norms are stabilization or standardization processes. They are systems for facilitating commitment of people to actions supporting obligatory boundary maintenance.

Logic is a system which discovers and stabilizes the structure of possible relationships among symbols. Logic seeks to eliminate the unessential, the impossible and improbable from any verbal or mathematical statement. Semantics is a system for stabilizing the meaning of word symbols. Human speech and writing are instruments for communication which have their stabilizing or informative function (their model or standardizing effects), just as they may have their transformational or complicating effects.

The presuppositions of scientific inquiry include recognition of the function and value of stabilization and simplification in system processes. Science assumes that the universe is not immeasurably capricious. Its faith or method of proof requires that human observation and meaning are reinforced only through redundancy, through public rather than private inspection, through the exact replication and reconstruction of events. The scientific method and its techniques of prediction are designs for stabilizing all or most of the environmental boundaries except those deliberately transformed in order to measure the effect of a specified induction.

The electronic computer is an additional device invented by man to both stabilize and transform his environment. It was a successive and rational step in human efforts to both carefully advance and then stabilize and control the environment. Like previous human inventions, the computer was designed to extend man's environmental control or power, hopefully accomplished at the same time by a reduction in human effort. These laws of simplification or economy, of cost-benefit ratios, regulate to a considerable extent the direction and extent of all planned human change.

The effect of cybernetic or information technology upon human behavior is becoming clearer with increased operational experience and understanding. It appears that electronic computers and other information processing machines possess these particular characteristics:

1. Tremendous speed in all forms of calculation—all arithmetic operations and logical processes of comparing, sorting, and deciding.
2. Perfect accuracy in all calculations and perfect nonbias in logical processes.
3. Almost constant and continuous availability for work and total attention to the focal problem(s). There are no labor problems or personal distractions. They are complete slaves.
4. A vast memory storage capacity for unrelated information bits.
5. Rapid and certain information retrieval (an infallible memory).

Mechanized information processes seem to have the following general effects when applied to well-defined routines (problem-solutions):

1. Humans are further relieved of many repetitive processes involving routine mental and physical skills. Human attention in mechanized information systems tends to change from continuous surface attention or supervision to irregular attention to operations by supervisors and/or concentration by planners and programers.
2. Computer control technology permits error detection and response before many systems problems became serious (serious enough to be detected by human observation or observation by persons external to the operating system). This control capability prevents significant losses in time, materials, energy, and money. It is a substantial cost reduction factor.
3. Information technology systems further increase efficiency by speeding data processing, especially numerical and logical processes.
4. Information technology permits penetration into very complex, expensive, and abstract inquiry problem areas.
5. Economic patterns are affected by automation. Direct production costs are often reduced. Indirect costs for planning often increase, at least temporarily. Planning and programing goals tend to become more long-range. Demand for intellectual skills among workers has increased.
6. Human technology is expediting the rate of change in methodology, in theory development, and in the accepted order of human values.
7. The inquiry process has assumed a higher priority in our technological society. Its parameters of application are continually expanding. Man is aware that he has greater ability to shape his future experiences than he once imagined. This means that his control power is expanding; it also means that his new-found capabilities will have to be used responsibly.

8. There is an increasing educational and communications gap between those who understand and exercise the power and limits of systematic inquiry and those who recognize neither its power nor its present limits.

Information technology has flourished in areas of human endeavor in which the tasks to be performed are routine and systematized and where there is a history of experience and know-how. Automated control of manufacture and distribution and business data processing has revolutionized these processes. Computers and computer languages have been designed for both business and inquiry purposes. Extensive libraries of software tapes have been developed which extend the use of computers into solving either a general class of problems or a specific problem.

Modern business education includes training in operations research and in mathematical and logical processes applicable to the prediction and control of the economic environment of business. Scientific business methods are replacing many rule-of-thumb operations in American business. Relatively current and complete information is more readily available for business decisions, whereas the same decisions would formerly be almost wholly intuited.

In addition to being associated with control function, cybernetic information technology has been used in connection with extremely complex human inquiry, focusing on problems requiring rapid and accurate calculation. Scientific and mathematical computers are those with the greatest capacity and speed. Governmental and university research centers are the operators of most of the major scientific computers.

Mathematical and computing techniques useful for making programed decisions have replaced or supplemented men in business and in science, but computers have not been equally successful in simulating man. Cybernetic theorists are very much interested in the simulation of human behavior on a broader scale, and behavioral scientists are interested in the theories of human behavior emerging from these research efforts in human simulation.

Computers which simulate individuals or organizations often use heuristic programs or humanoid problem-solving or search techniques rather than the algorithms of classical mathematics. Heuristic programs incorporate alternative mapping or branching processes which are designed to advance the process of solving or to resolve diverse and ill-structured problems. They even can be programed to act unpredictably and adaptively, but they are far from being both autonomous and adaptive in respect to their environment, as are men.

A number of cybernetic theorists believe that machines will be able to simulate most human inquiry processes in the relatively near future. (Simon, in Pigors, pp. 86–101) They predict that human "mental flexibility" will be simulated long before human sensory-manipulative flexibility. If this is true, man, in his future man-machine interfaces, will maintain his major dominance in systematic operations involving general adaptability, physical mobility, and sensory and manipulative capabilities. He will become something of a problem generator, a goal inventor, a creator of consumption needs. The technological information system will be the problem solver and the producer.

The full meaning of the cybercultural revolution is at present not at all well-understood. System philosophers do know that there is a substantial qualitative difference in the order of problems that confront civilizations possessing extensive technology. There is considerable consensus that social changes will be accelerated to the degree that men must not delay in speculating about inquiring into and ordering their value priorities.

System experts agree that the identification of human issues, goals, and problems becomes much more important as the means or alternatives for their attainment is assured. Initiation of social acts without assessing action values will become an even more terrifying thing than it has been in the past. The central problem of future societies will be to identify and understand social goals in order to optimally exploit the new means.

The implications of system philosophy, psycholinguistic and cybernetic theory, and technology relative to social ideology is abundantly clear to advanced nations. The power of men to exploit their natural resources and human intellectual powers is greatly increased. Men will be freed from the drudgery of repetitive mental work just as they are being freed from physical labor.

It is believed that the computer coupled with production machinery will inevitably exceed the social capacity to create production-oriented jobs. Without substantial changes in the social production-consumption system, fewer and fewer people and nations will work harder and harder to control the inevitable centralizing tendency which accompanies technological systematization, while more and more people will be disengaged from the production of essential goods for greater periods of time.

For the majority of persons, the problems of leisure will be the major concern in an automated production society. The psychological importance of such an alienation of men from the rather obvious purposes and motivations that accompany productive work is not well understood. We are beginning to see in the United States that this type of social alienation may contribute to an explosive social condition.

There is consensus among system philosophers that men cannot continue to be as exclusively self-interested and utilitarian in their social orientation as they have been in the past. Nor can they allow their technology or economy to completely dominate and shape their individual and societal futures. System philosophers believe that the minimum requirement for the survival of the human race will be the extensive development of an individual-social responsibility sense that is greater than utilitarianism. It will require a more abstract or higher value system. Although the history of human civilization does not disclose so clearly the positive effects of social cooperation as compared to technological advances, the major human goal of the future will be to resolve this very matter. The determination of individually and socially acceptable human values and systems of values will be the major problem emerging from the cybercultural revolution.

21

The Designing and Programing of
Instructional Units

INTRODUCTION

Probably the reader is by now fully aware that the very essence of any system planning process is its organismic nature. System planning and programing begin with a statement inclusively defining the total system or program under consideration. This statement defines the anticipated final state of the system as a whole at the completion of its proposed operations. Then the process of system analysis and redesign proceeds through an orderly series of steps involving first macroanalysis or abstract planning and then microanalysis, the specification and materialization of a real microcosm or exemplary prototype. The programing or design phase involves the specification and synthesis of the prototype and the eventual integration of the new model into the full scale operation of the system.

In instructional programing the ultimate purpose of the system, the desired service provided the learner together with his desired response, is at first imagined, described, and specified in terms that are simultaneously idealistic and attainable. Next the major subprograms, objectives, and learning behaviors are specified. Analysis then proceeds to identify in a recursive and decomposing manner the sub-functions, tasks, activities, and learning content or facts involved in achieving the learning objectives. When actually implemented, the system reverses the process once again and operates progressively. It follows a well-ordered sequence of events which cumulatively modify the learner, changing him in successive increments from his present state into the predetermined program termination state.

In the process of instructional program analysis, care must be taken to account for all necessary and sufficient learning components. Analysis involves determining a hierarchical or taxonomic schema where the total of all subclass goals or learning behaviors specified represents an exhaustion of the universe set of all critical behaviors needed to assure the accomplishment of the final or incremental goal, a more precious ultimate state.

An effective and efficient linear or straight path through the learning program is a desired goal of programing, but in complex learning programs the experiences provided almost always permit some deviations, reversals, repetitions, and overlaps which reduce the function and therefore the necessity of absolute linear sequencing. Thus the planning of programs as well as the actual experiencing of learning programs often make provision for heuristic or search strategies and tactics to modify the necessity of absolute linearity.

Whether an educational design problem involves the planning of a complete curriculum, a course of study for a building unit or grade, or an abbreviated unit of study, the programing procedures are approximately the same. Learning program planning, stated simply, includes:

1. The setting of specific performance objectives, the final output states or integrated sum of a sequence of modifying events.
2. The application of logical analysis in the determination of learning content units, the response behavior components, and the ordering of these units into a desirable sequence.
3. The specification and material development of real content items and methods-means-media vehicles (input resources and experiences) fully representative of specified logical and theoretical components.
4. The vigorous measurement of the results of actual experiences in comparison with the specified theoretical performance objectives.

The particular schema used in the initial planning of a comprehensive curriculum program is a matter of arbitrary choice. In instructional program design, planning is a point of beginning, not ending. The system principle of equifinality applies here. Whether subject matter, social-institutional needs, or the learners' psychological requirements are the schema originally chosen for planning is a matter of relative unimportance; the problems are the same. All of these factors must be accommodated and integrated into the actual learning operations. Technology and methodology always involve concepts, materials, human interactions, and organization.

The important choices in original planning require that, first, whatever schema is chosen, it must be considered as a complete or total system or universe—it must exhaust the universe set. Secondly, the planning must then be directed toward selection and simplification. Through the exercise of careful theoretical modeling and systematic judgments among a rich repertoire of alternatives, planning can pretty well determine probable workability and economy. Finally, thorough planning provides for the operationalization and evaluation of the program, determining in advance the criteria to be employed in assessing act and action quality and meaning.

At this time readers are referred back to the MARS model of system analysis and redesign which was presented earlier. They will find that the designing of instructional programs or learning units is similar to designing any complex operation or system.

The output of a well-designed learning program is a documented instrument useful in determining both before and after the fact the validity and reliability of instructional objectives, content, sequence, method, media, and degree of achievement.

Optimum educational programing should carefully apply the logical and scientific principles of selectivity, simplification, and workability and the additional system principles of economy and social utility. Learning is an investment. Rational judgment of a learning program's efficiency must be made on some comprehensive cost-benefit analysis, comparing alternatives. The total of all resources committed to a program, including learner and instructor time, are to be considered in comparative economic terms, for they are all scarce economic resources. As education is a deliberate and costly human operation, it should accelerate the attainments of the learner and indirectly benefit the client system or social environment to some degree greater than nonsystematic, natural, or random growth would attain. A careful consideration of this fact points out the need to recognize that formal education is indeed specialized, even if comprehensive; it is only a part of the total of all worthwhile experiences a human being undergoes in living.

If educators acknowledge the need for an economical selection of learning experiences to be included in a specified instructional program, their planning objectives will be chosen according to these operational principles. It will contain:

1. The minimum specific skills, knowledge, and attitudes needed for group and individual living, appropriate to the maturation of the learner.
2. The minimum reading, writing, speaking, listening, arithmetic, and thinking skills.
3. The minimum specific perceptual and psychomotor skills, knowl-

edges, and attitudes needed for entry level and initial success in practical or vocational living or for continued success at the next higher level of education.

4. Generalized physical, social, academic, and vocational skills, knowledge, and attitudes for enriching the learner's future life and the environment in which he lives.

As a learning program involves multiple goals and therefore multiples of input resources, content, and experiences in such a way that it facilitates successful cumulative modifiability, a systematic program must weave both the learning event inputs and the learner's responses into the very fabric of the instructional plan. Instruction and learning thus become a two-way communications process, exercising the multiple loop cybernetic networks of the learner and the instructional system independently as well as dependently. An adequate instructional system provides for the best possible continuous, complete, and concurrent feedback to both the learner as receiver and the instructional system as source (the teacher, computer, teaching machine, curriculum author, etc.).

The application of system principles to instructional programing now has a historical experiential background which provides the general educational practitioner with a systematic theory and operational technology possessing almost universal applicability to the planning and implementing of new and different curriculum units. These tested and standardized procedures include the following sequence of programing tasks:

1. Establish program goals; identify and analyze necessary input components, consider alternatives, conduct interdisciplinary feasibility analysis, and select the optimum plan from among preconfirmed technologically workable, economically feasible, and socially beneficial alternatives (see MARS model, Steps 1–8).

2. Develop a prototype model or unit of an actual learning program and a representative criterion test measuring all of the expected performance standards. Both of these programs employ experiences and materials representative of the learning objectives stated earlier (MARS, Step 9).

3. Validate the criterion test, representative of all behavioral outputs expected in the learning program by giving it to trained and untrained sample populations of learners. The majority of untrained learners should fail the performance test items, responding incorrectly. Conversely, the majority of the trained or expert population (85 percent) should respond correctly to the representative test items (MARS, Step 9).

4. Learning content and experiences are eliminated from the criterion test if the untrained population responds correctly to related test items, as they are unnecessary. If the criterion or expert group responds incorrectly to the criterion test items, the related objectives and experiences should be considered for elimination also, as they may not be relevant to the true goals of the program (MARS, Step 9).
5. On the basis of the criterion testing, either the test or the related program or its objectives are revised where necessary. Next the program itself is tested in a pilot or laboratory setting. Unnecessary or unproductive experiences are modified or eliminated. Alternative learning methods, means, and media are tentatively programed and tested. Learning sequences are established and corrected. The program is broken into functional practical units. Learning programers recommend that learning frames or response units and their larger practical units permit a minimum of 85 percent success in attaining the desired output proficiency (MARS, Step 10).
6. Following criterion testing and informal and formal laboratory testing the learning program is then tested in the field, further revised; general performance standards are established for field operations (MARS, Steps 11–12)
7. The necessary methods, means, and media vehicles are developed or purchased for putting the prototype model into full-scale production. Operational personnel are trained. The system is given a trial run. The system is then put into full operation (MARS, Steps 13–15).
8. Finally the program is integrated into all other components of the total instructional network, including all management control subsystems. It is revised periodically on the basis of continuous and concurrent reassessment of its effectiveness and efficiency. The program is now considered fully operational (MARS, Step 16).

Some elaboration of procedures for programing which are new to many educators will follow later in this unit.

INTERDISCIPLINARY THEORIES SYSTEMATIZING INSTRUCTIONAL TECHNOLOGY

In the programing and application of instructional technology the predefined contents (skills, knowledge, attitudes to be communicated) are developed through an instructional system or network which accommodates and integrates subject matter, the learner's psychological responses, and the technological means and media of presentation. This is done in a learning environment which imposes upon the functioning system input-output restrictions which are both social-political and economic in nature. Thus an instructional program is a system with both uncontrolled and self-controlled restrictions and constraints, as well as focal objectives.

Within the parameters determined by identifying acceptable program objectives, the optimum instructional program is a series of precisely controlled learning experiences which are organized into a linear chain of dynamic events which require the learner to respond and adapt while fulfilling the requirements of the task-relevant learning criteria in an efficient and practical manner.

An instructional program is both a product, a predetermined and complete structure, and a process, a mediating heuristic or search method or package. A well-planned instructional program is the result of combining past interdisciplinary theories and experiences into an empirical method or system capable of further refinement and improvement.

The optimum process and ideal product in instructional programing involve the discovery and/or invention of desired response behaviors, the specification of what it is that controls these behaviors, and the creation of appropriate conditions or techniques needed to shape these behaviors. Instructional programing predetermines:

1. Predetermines all skill, knowledge, and attitude components necessary in attaining the learning objectives. This predetermination of learner output components or responses identifies what the learner is to do as a result of his learning experience, under what conditions and limitations he will be expected to perform, and to what standard or level of performance his experience should be guided.
2. Predetermines the specifications of input components, the resources, technologies, and situational stimuli which challenge the learner to respond in the prescribed way.

In efficient learning programing the general pattern of learning experiences is progressive and incremental. Programed learning events are serially ordered or controlled inductions of paired and balanced concepts and response acts which contain neither more nor less learning content than is needed.

Cybernetic or system psychology requires that the learner must respond actively to confirm all of his motivational, conceptual, and skill learnings. As soon as possible he is required to make a meaningful overt response to instructional stimuli to confirm his learning. This need not be purely a complete reconstruction of a verbal input or physical demonstration; but it might be just this in the early stages of the learning sequence. The overt response of the learner provides an opportunity for feedback to the instructor and learner cognitive control systems, which, in turn, may facilitate communication and refinement of the act and the assessment of its value.

Learning feedback, either self-mediated or mediated by the instructor, is most useful when it facilitates immediate confirmation. Only then does it permit immediate stabilization or modification, the ideal condition required for any high level of learning efficiency.

The act of instruction is designed to manage positive reinforcement or confirmation. Instruction should control events to the degree that they function as a series of mostly successful experiences, generally producing positive rather than negative reinforcements. In order to achieve positive reinforcement of learning events and sequential or cumulative success, the events must be organized into a chain which facilitates the optimum rate or pace of successful learner input-output transformations. Ideally most learning experiences are individually paced. However, in many instances group pacing is necessary and desirable. Even in programs involving primarily individual pacing the instructor generally establishes periodic deadlines and milestones which improve the learner's performance over his natural or unsystematic pace. Otherwise the formal learning system would have no justification.

Optimum learning programing is designed to assure that the learner gets only what he needs to know only when he needs it. The information he receives should be primarily appropriate to his needs and abilities of the moment; only secondarily does he acquire a reservoir of learning inputs for immediate storage and indirect or long-range (investment) benefit.

If a learner does not learn when he is committed and subjected to the learning system, it is not his fault but that of the functional system constraining his experiences or the environment restricting the instructional system.

In a learning program, material and tasks are organized to allow the learner to pick his own best route to the desired terminal behaviors. If need be, he may detour from the mainstream into a remedial or tutoring program entirely free from a feeling of failure. The best instructional system keeps all negative reinforcement to a minimum, but, if it accommodates negative reinforcement, it seeks to ascertain that such reinforcement is identified first in the situation, secondly, in the learner's feedback system, and only as a last resort, in the individual mind of the instructor.

Expressions of negative reinforcement should be moderated by the instructor as representative of inappropriate input-output responses, as special situational conditions which are restrictive. The instructional system then encourages the learner to initiate heuristic responses which by-pass the original restrictive barrier if possible and practical.

Any learner should be permitted to by-pass material and interim steps in a programed learning sequence any time he can demonstrate the terminal behavior to the degree of proficiency established in the planning criteria. Actual learning content should be determined largely by the learner's individual needs and response patterns.

A learning system is validated by ascertaining whether it does what it is supposed to do. No system should be held fully accountable for accomplishing anything that was not previously specified as an output objective. Yet an instructional program should always seek external as well as internal validation. It must ask whether it serves as an instrument for meeting the ill-defined educational needs of the target population as well as the prescribed objectives. Only in this way can it become an instrument for long-range social advancement.

The psychology of learning has hypothesized, tested and confirmed the existence of seven factors or input conditions (OEO, pp. 5–12) which appear to generally affect or stimulate productive learning. The first of these factors is *motivation*. A learner must want or need to learn. To some extent self-motivation is both prior to and after the fact of imposed motivation. A learner must want to and need to learn and he must be aware of this fact. Self-motivation is to some extent an uncontrolled instructional variable; yet instructional techniques do exist which indirectly stimulate or control motivation. Instructional programing may control motivation by:

> Presenting learning experiences which are intrinsically challenging, where each learner immediately generates his own incentive.
>
> Setting distinct and realistic goals.
>
> Presenting experiences which allow the learner to confirm via his own feedback system the fact of success.
>
> Pointing out the utility and early use of the learning response.
>
> Informing the learner exactly what he can do with the new knowledge or skill.
>
> Demonstrating for the learner or persuading him that the learning goal is desirable and worth the inquiry or learning effort.
>
> Convincing the learner that there is a chance for success; he must feel that the goal is attainable.
>
> Providing supplementary praise where it appears to be needed.
>
> Frequently instituting exams, quizzes, or problem situations and challenges for sustaining learner motivation.

Presenting occasional problems which result in a degree of failure, balancing the normal pattern of general and/or total success.

A second factor of significance in instructional programing is *organization.* Learning experiences must be perceived as meaningful to the learner if they are to be valued and voluntarily reconstructed. The acquisition of an understanding of relationships of cause and effect or rules of procedure among learning content elements is necessary to assure meaning. An instructional system induces organization, meaning and integration into learning in the following ways:

1. Learning is enhanced when the instructor provides a preliminary overview or introduction to the material or process to be learned. An adequate introduction communicates in a code familiar and appealing to the learner. Introductions relate the new material to the learner's past experiences.
2. Demonstration and simulation assist learning organization and integration. Only the essentials of a task should be stressed during a demonstration or simulation. Too much detail too early may be harmful in instruction. Extensive didactic or deductive explanations will be more useful as the learner becomes more interested and expert in the general area of learning.
3. A final review or summary of a complex learning experience or series of experiences is integrating and organizing. A lesson is often productively closed with a review in which the instructor tells the learner what they have learned.
4. Learning organization and integration are accomplished through program patterning. Inductive learning patterns often provide the best organization for learning. In inductive instruction the learning stimulus is first presented as an integrated problem, complex act, or observed demonstration. Then it is broken or decomposed into a time and space ordered series of subfunctions. Finally the learner is encouraged to synthesize the subfunctions into a concluding experience confirming the learning gains directly and completely. Careful preprograming and heuristic trial of learning event sequences determine the most natural or best order of space-time-function among learning activities.

Another significant factor influencing learning is *participation.* Programed learning is based upon an assumption that an individual learns only by becoming directly involved in the system—by making both a mental and physical response which the learner initiates and organizes (via feedback) himself. Participation itself both motivates and organizes. Mental attention to lectures is maintained best by interspersing straight information reception with opportunities to respond or to initiate expressions of understanding, application and evaluation. Verbal responses are best if the learner actually writes or recites his responses and makes them public.

In teaching through demonstration, interruptions and pauses should be inserted into the learning program also, in order to permit and require the practicing of each activity. In lieu of actual practice, demonstrators might encourage a learner to visualize or imagine the feeling of the learning task under consideration. Another alternative to actual practice is involving the learner in a reformulation in his own words (a description or explanation) of the process.

Learning of difficult, critical, and complex responses cannot be accomplished without participation. Motivation and organized presentation are sufficient only superficially. Higher levels of skill, knowledge, and attitude acquisition require both overt verbal and physical participation.

A fourth factor or condition of learning is *confirmation* or *reinforcement*. Learning must be checked and tested in the mind of the learner. He must compare his expectations with what actually occurs in order to determine first if he was right (confirm his hypothetical meaning) and second why he was right (confirm his evidential meaning). An instructor initially demonstrates right ways only and encourages positive confirmation and reinforcement. Only later are erroneous demonstrations provided as negative reinforcers to assist in defining the functional parameters of the learning responses.

Learning is materially confirmed when a learner demonstrates that he *can do* something he could not do before. It is maximally confirmed when a learner *does do* something voluntarily in an appropriate setting. This latter type of confirmation involves or demonstrates an integration of the conceptual, psychomotor, and affective domains. Proper predispositions are reinforcing proper dispositions or capabilities in this ultimate demonstration of learning.

A fifth factor or condition affecting learning is *repetition* or *reconstruction*. Repetition involves the technological reconstruction of events. Accurate reconstruction or replication is most useful in facilitating the acceleration of skill acquisitions. Repetition has little to do with reinforcing cognitive retention. Cognitive retention is better reinforced by purposeful and meaningful practice which overcomes the demotivation effect generally accompanying drill (absolute repetition).

The best criteria for justifying repetition are found in the significance or complexity of the learning tasks. Are they crucial, difficult and complex? Do they require much internal patterning? If they are, the best pattern of learning is probably an inductive pattern, a whole-part-whole learning sequence. Often teacher demonstration, pupil demonstration, mutual analysis, step-by-step practice, and finally reintegrated pupil demonstration or reconstruction plus an integrated review or summarization by the instructor provide the best pattern of repetition of significant or complex learning programs.

Factor number six in determining effective learning is *application.* Application is an early step toward generalization of practice in a meaningful and utilitarian manner. It induces an attitude or set toward evaluation and valuing. It facilitates discrimination and meaningfulness by providing an opportunity to discover the limits of appropriate and productive application. An instructor can guide learning transfer by demonstration of applications of learning to new situations, using both positive and negative models.

The final factor expediting programed learning is *problem solving.* The application of learned responses and rules to the solving of novel or complex problems involves the combination of all previously applied factors. Skills, knowledge, and attitudes are combined in a meaningful set of rules to create an even higher order system of rules for decision making (rules of evaluation and valuation). Demonstrations are helpful in encouraging the learning in productive problem solving. Again, the cruciality, complexity, and costliness of the learning process should be the primary criterion for determining the need for guidance in problem solving.

The above seven factors or input conditions for guiding learning (motivation, organization, participation, confirmation, repetition, application, and problem solving stimuli) must be articulated with the several types of learning responses possible if the learning program is to be maximally effective. Robert Gagne and other behavioral scientists (OEO, pp. 22–35) suggest that a systematic taxonomy of overt human response categories to specific learning stimuli may include:

1. Specific responses—singular verbal, psychomotor, or affective responses to specific stimuli.
2. Motor chaining—responses involving a sequence of linked psychomotor responses necessary for achieving a unitary action or product.
3. Verbal chaining—production of a sequence of words which result in a unitary meaning (verbally represent an act or input-output relationship).
4. Discrimination—making different identifying responses to two or more stimuli.
5. Classification—making a response which identifies or categorizes an entire class of objects or events.
6. Rule using—Performing tasks or activities according to certain organizing rules of procedure or principles.
7. Problem solving—solving a novel problem by combining previ-

ously learned rules and stimuli to create a higher order rule response or generalization system. Problem solving involves both inductive and deductive reasoning and demonstrates the articulation of the determinate level of meaning with prior levels of hypothetical and evidential meaning.

In the systematic articulation of input stimuli conditions with specified levels of desired output response, an instructional programer can employ the following tested principles in his planning. Learning specific responses usually requires participation, confirmation, and repetition. Motor chaining involves participation, confirmation, and repetition. Verbal chaining requires these three factors supplemented by demonstration, verbal cuing, and discussion. Discrimination, the contrasting of different responses to two or more stimuli, involves the early levels of learning plus contrasting practice. Classification requires early levels of learning plus organization, confirmation, application, and discrimination.

Rule using, a higher order learning response corresponding to the level of cognitive synthesis in Bloom's taxonomy of cognitive objectives, requires the additional production of a set of operations and the derivation of a set of abstract relations. The equivalent level of advancement in the affective domain is perhaps the organization of a value system. (Krathwohl, pp. 176–193)

Actual practice is generally necessary in developing skills for using rules, although adults may sometimes learn new principles vicariously— entirely from a verbal statement of that principle. Repetition is of little value in rule using. In guiding the acquisition of skill in rule using, an instructor encourages the learner to discover relationships among two or more concepts. He may ask for verbal descriptions or require demonstrations and accompanying explanations. He may request verbal statements of subprinciples. The learner may be led by a direct series of verbal cues to formulate rule statements himself, which he later confirms in application. The culmination of rule-using responses is demonstrated when the learner develops a complete, refined, and standardized (fully acceptable) formal statement of the rule which reflects its full and true meaning.

The ultimate learner response is of course equivalent to the resolution of the ultimate learning stimuli. A problem is identified and productively solved, and all of the accompanying meaning is understood. The learner is in command of the task and of its evaluation and valuation.

Before concluding this brief discussion of the systematic applications of learning theory to instructional technology, it seems desirable to re-emphasize that programed learning involves a relatively closed or systematic linear or incremental order of selected learning input stimuli and output responses. Systematic psychologists and instructional programers perceive affective learning to progress from simple and immediate wholes to complex and abstract wholes and higher order systems. Gagne, Bloom, Krathwohl, and other systematic learning theorists perceive that learning becomes progressively more difficult if any early stages in the learning input-output chain are omitted completely.

All educators must acknowledge the fact of individual differences, however. Some learners and some learning stimuli exercise intuitive learning—complex simultaneous or gap-leaping responses. A teacher may guide intuitive learning and intuitive learners differently from systematic learners, although alternative emphasis on inductive and deductive methods of presentation, psychomotor-affective involvement in addition to cognitive involvement, and active learner demonstration and explanation should adapt instruction to either the intuitive or the systematic learner's needs.

A truly comprehensive learning program will include both intuitive and organized learning stimuli, however. In social value instruction and instruction in creativity and inquiry, learning experiences as presented are deliberately complex gap-leaping, significant, and impact laden. Other learning experiences require deliberate simplification, ordering and objectification. The extremely complex academic cognitive areas of language, mathematics, and the natural sciences are quite adaptable for the most part to atomistic or systematic learning. Economics is perhaps that social discipline which is most adaptable to programing; politics, sociology, history, and the fine arts and humanities require considerable inductive learning to be truly meaningful.

Finally, the learning programer must recognize that there are rather determinate properties in learners which may be beyond the control of any instructional program. There are the persistent properties of general ability, of available intelligence, creativity, and psychomotor skill. There are conditions of maturation that definitively restrict the rate of educational development. Aptitudes, mechanical, spatial, and verbal perceptual speed and kinesthetic sensitivity may vary. Previous knowledge of requisite cognitive specifics may be lacking. Learners differ in their facility for learning through the eye or ear or by multi-media.

Learners have different persisting and varying patterns of attitude and interest. They may have a different history of experience with various teaching technologies (lectures, demonstrations, films, direct manipulation, opportunity for practice, application, problem solving, exercising judgment, etc.).

Jerome Bruner's idea that basic concepts in any discipline can be taught in some meaningful way at any age or level of maturation or development serves as a useful stimulus for planning for the inclusion of significant concepts and experiences in all learning programs and for encouraging selectivity of both content and experiences rather than coverage of detail. The idea even encourages comprehensive planning in an indirect manner. However, this statement per se does not serve to conceptually determine many of the systematic variables involved in effective instructional programing. It does not fuse the desired or imagined output goal to the present conditions of the learner in a normatively probabilistic, real, or practical way. It does not advance the design of a program which actually determines what is accomplished. Programing requires determining the desired output state(s), together with all essential input conditions, including the restricting environmental and constraining system parameters; only when this is done can an instructor establish an efficient system or technology for answering the instrumental or "how" question, the "when, where, and how much" questions—all objective and practical questions which delineate the actual fact or possibility.

THE DESIGN OF LEARNING PROGRAMS

The design phase in learning programing begins the resynthesis of prior analytic accomplishments. It does this by giving material substance or body to the abstract plan or model. The actual learning program is a real microcosm.

Programing design at first involves the extension or transformation of the prior systematic macroanalysis as demonstrated in the MARS model, Steps 1–8. It requires a one-to-one transformation of idea inputs into representative material means and method inputs. Systematic design at first controls the error possible in this transformation. When the transformation into reality is accomplished, however, instructional programing begins to simplify and synthesize the material inputs into an integrated operational reality.

The second major phase of instructional system analysis begins with a review of the macroanalytic plan or model (the product of MARS, Step 8). Then it proceeds to design a prototype model of means, methods, and media which veridically and faithfully represent the plan (MARS, Step 9). Prototype planning involves microanalysis, a task and means-method analysis detailed enough to guarantee a step-by-step progression of a series of actions leading to the ultimate output condition previously established.

Resources for selected representative learning content (learning concepts and behaviors) are the skill requirements demonstrated by technologically working systems, including expert human operators. In task analysis the programer breaks down output goals and general plans into ordered material inputs and actions. In the process of designing a prototype learning program a designer first analyzes the task or activity content, concretely identifying:

1. The cues, signals, indications, directions, that call for action or reaction.
2. The control input objects, tools, or content to be used or manipulated in the learning experience.
3. The action or manipulation to be made.
4. The cues, signals, and feedback which indicate that the action taken is or is not correct and adequate.

Learning plans, learner intuitions, instructor exhortations, and learner reactions to problem situations all provide initial learning action cues. The heuristic or searching aspects of the learning event involve the selection of input tools or resources to be manipulated and the manipulations to be exercised. The final feedback experience involves both an intuitive and organized mixture of cues confirming all of the elements of meaning of the act and action.

In the selection of tasks for program design and implementation a knowledge of task identities is the minimum information needed for performance of the learning event. Knowledge and understanding of alternate means/methods is more adequate information. Know-how achieved through experience in discrimination, classification, application, and evaluation requires more advanced or higher levels of preparation for ascertaining the adequate selection of learning tasks. In instructional programing, the planning theorist in all probability must be assisted by experienced practitioners, both teachers and learners, in the task selection phase and in all subsequent steps in completing the program design.

Task selection for prototype programing answers these questions:

1. What does the learner need to know and do (stated as response behavior)?
2. What are the conditions under which the learner will perform (the learning context, content, means, and media employed)?
3. What are the expected performance standards (these must be systematically representative or normative)?
4. How can the response behaviors be observed and measured?

After one or more actual concrete learning tasks and response behaviors are identified as representative of each learning objective, the educational programer begins to order his tasks into incremental lessons or units. Normally the ordering of instructional activities is a relatively heuristic or intuitive experience. This is particularly the case when planning is unsystematic and when prior teaching and learning experience is lacking.

Typical curriculum objectives and courses of study in schools are general and vague to some extent. Often teacher strategies are purposefully flexible and adaptable. Usually learning strategies and tactics vary. This is and has been the norm for most instruction mediated by human instructors and the customary instructional materials. Only instruction in complex developmental subjects in schools, those which involve considerable psychomotor and academic chaining, tend to be presented in a more orderly fashion.

The precise and penetrating programing of a computer and its related technology is still more systematic, however. Thus the knowledge gained through experiences with system technology and its derivative theories is a potential source of information for any educator who wishes to extend his penetration and proficiency in the skills of planning, programing, and evaluating instruction, including his own technology or methodology. There are two programing strategies which have been developed by instructional technologists which appear to have the greatest utility for broader application to formal education and instruction. Paragraphs below will explain criterion testing, which is useful for improving program content selection and validation, and a matrix technique for determining the optimum sequence in learning events, improving program reliability.

CRITERION TESTING

When a prototype instructional program is designed and made ready for operation, there is still much uncertainty remaining as to whether its content is irrelevant and therefore invalid and whether it is organized in the most efficient manner. In most instruction in schools today the answers to these questions are intuited, assumed, or unconfirmed. System analysts and programers have developed more precise means for predetermining the answers to these questions and measuring them after the fact.

Instructional system programers assume that instructional technologies are tested by applying measurement techniques which are criterion-referenced, comprehensive, valid, reliable, objective, standardized, and economical. A carefully designed prototype learning program systematically meets the above performance criteria. The addition of criterion-referenced tests to tests of performance or proficiency tests is a singular contribution of system theory. Criterion-referenced tests add absolute criteria to the measurement of learning. Tests of proficiency only establish relative criteria.

When a prototype learning program is first designed it needs to be validated. A special criterion test should be constructed to serve as a quality control instrument for program validation.

A criterion test is a comprehensive and proportionate representation or blueprint of a prototype or model learning program, almost an alternate form of the program or its learning "items." It is based primarily or originally on the prototype program previously designed, which, in turn, is a reflection of the care and accuracy built into the program task analysis. Of course the prototype program and its elements are dependent upon prior system analysis for their selection and justification.

As a criterion test is not given to persons experiencing the actual model training sequence, it is a method for determining the external validity of the program. And it must reliably represent all of the significant knowledge and behavioral responses expected from experiencing the pilot program if it validates that program.

The validity of a criterion referenced test is determined by giving the test composed of representative program input-output response items to two target populations. First, a group of persons who are recent confirmed experts in the expected learning are tested. Next a group representative of the learners for whom the program is designed are tested. Test validity is determined by measuring the difference or discrimination between these two criterion groups indicated by each item and the test as a whole. At least 85 percent of the trained population should respond correctly to the items in a criterion test, while a similar proportion of the untrained group should generally fail to respond correctly. If most of the experts fail the items, there is a question whether the learning experiences they represent are truly relevant, valid, and necessary. If both groups generally respond correctly to the items, the items or instructional tasks are too easy, another indication of probable irrelevance. Thus criterion testing is effective in setting both functional upper and lower limits for course content. It is a useful tool for predetermining the systematic qualities of functional appropriateness and economy in the criterion instrument first, and then the knowledge gained in this process is transferred into a modification of the prototype program.

The criterion test is considered reliable when it accurately and consistently measures the combined performance of the two criterion groups. The usual measures of test reliability can be applied in determining criterion test reliability.

It should be noted that criterion test items are derived as directly as possible from the prototype training program. Therefore the difficulty of test items should not be changed if an item does not function as desired. Rather the item remains the same even if the wording or structure of the experience is modified or improved. If criterion test items do not function appropriately, the learning program should be reviewed and modified. Validated criterion tests should serve as agents of both stability (standardization of the program) and change (modification of the program). Thus when the prototype program is heuristically modified and ultimately determined by a series of criterion tests, the final version of the criterion test must be consistent with the final version of the prototype program.

In addition to the development of criterion tests for validating the program prior to its standardization, and subsequent laboratory and field testing, criterion tests serve a useful function in facilitating and assessing a learner's performance as he progresses through a systematic learning program. Other criterion tests are designed to provide a feedback of:

1. Immediate and continuous within-lesson assessment of *act* effects (by both the learner and instructor, where possible).
2. Immediate and continuous within-lesson assessment of *action* effects (by both the learner and the instructor).
3. End of the lesson or end of the unit assessment in order to predict ability to proceed in the program.
4. End of the course assessment to predict retention and transfer of knowledge in application and problem solving and predict ability to proceed in the next learning program.

All of the above tests are criterion-referenced when they result in an 85 percent correct response to each item and to the test as a whole.

In developing criterion tests, care should be taken to assure that the tests remain truly valid and reliable measures of the instructional program. For this reason this type of test should require a range of behavioral performance measures. Performance measures and demonstration-explanation items should be employed in addition to paper and pencil items. As paper and pencil test items and performance items are designed to be valid, reliable, representatively comprehensive, and economical, they will probably be quite specific and objective in nature. Essay tests meet these criteria primarily if the ability to explain in writing is a goal of the instructional program.

If it is necessary to measure performance within learning groups, to distinguish between "good and bad" learners, norm-referencing procedures and test items should be developed for the program also. In such a situation different criteria are employed to select norm-referenced or relative items for measurement than are used to select criterion or absolute test items. Norm-referenced items are deliberately alternate rather than redundant learning experiences. They are typically selected so that they present greater difficulty. The appropriate average difficulty index for a norm-referenced test is an assumed 50 percent or normal success. Item analysis of norm-referenced tests is based not upon the expert and untrained criterion populations, but upon populations of "bright and dull" learners concurrently experiencing the learning program. Within the criterion-referenced instructional concept, there are no good and bad learners, only slower or faster ones.

The reader probably needs no reminder that the usual means of assessing validity and reliability should be applied to both criterion-referenced and norm-referenced testing procedures. Assessment of construct validity (logical or theoretical validity), face validity (means/method or technological validity), and both concurrent and predictive validity is desirable if the prototype learning programs are to be standardized and widely disseminated.

INSTRUCTIONAL SEQUENCING BY A MATRIX METHOD

It is probable that the theories and techniques of criterion testing will be valuable if applied to education and instruction more extensively. Like other system procedures, they can be expected to have their greatest applicative value when applied to learning problems and programs which are critical, complex, and costly, for it is then that induced precision, organization, standardization, and economization will have the greatest pay-off.

There is another system technique which can contribute to the quality of instructional programing also: The systematic determination of optimum job/task sequence by the matrix method improves internal or functional program validity and reliability.

It has been mentioned numerous times that systematic psychologists and learning programers perceive the normal order of learning behavior to be incremental, proceeding from simple, concrete, lower-level learning task units to more complex, comprehensive, and abstract tasks. Learning taxonomies reflect this order of development also. At this time a matrix technique for establishing optimum sequencing in a limited chain of learning events will be described.

Systematic job/task sequencing can be organized through the use of a simple matrix technique. The steps in the procedure are as follows:

1. Prepare a random list of terms, concepts, rules, operations or tasks,

motor chains, or problems as the learning objective. For simplification the use of concepts is suggested. Each learning unit should be written as a simple statement of fact (only one unique idea per statement). The statements are then numbered in the hypothesized order of presentation. Ideally the instructional inputs presented will each require a response in the actual learning situation.
Example:

1. Symbols represent something else.
2. Codes are symbols with meaning.
3. Symbols function as codes.
4. Some symbols represent objects.
5. Some symbols represent other symbols.
6. Books contain written codes representing spoken codes.
7. Maps employ symbol codes in a way different from books.
8. Some symbols represent spatial codes such as distance (a construct).
9. Maps convey coded meaning about distances by direct proportionate spacing of symbols on a page.
10. Maps are visually coded spatial representations.

2. Next a matrix or chart containing rows and columns is prepared. A number of squares equal to the number of learning units in the sequence are placed in a diagonal across the page. Each square is numbered and represents the single learning unit of the same number.
Example:

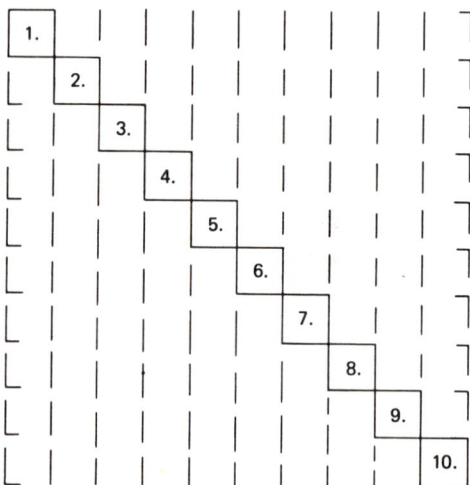

3. The next step involves indicating direct associations between facts by marking ×s in common squares of the matrix (above and to the right of the diagonal). Direct associations indicate common topics, central concepts, common object inputs, etc. All connections between learning inputs which are less direct are established indirectly in the learning chain or diagonal.

4. In the same manner significant differences in learning unit content are indicated in the common squares of the matrix below and to the left of the diagonal, by placing lines in these squares. Discriminations should indicate a high degree of apposition or distinction—different symbols, definitions, rules, central concepts, objects, etc. Thus a completed learning matrix indicates direct associations and differentiations by marking common squares. The only indirect associations indicated are in the diagonal (blank squares may indicate indirect associations also).

Example: A completed matrix of the above ten learning units

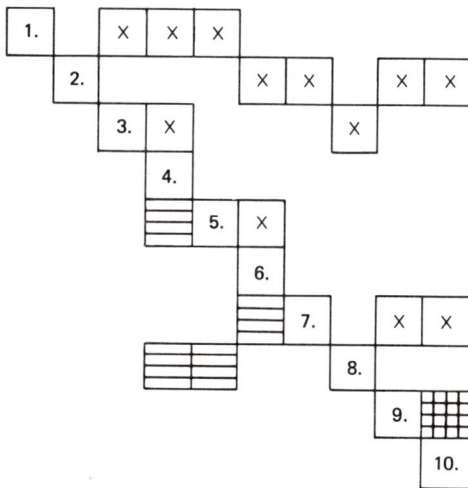

	1.		X	X	X					
		2.			X	X		X	X	
			3.	X			X			
				4.						
			≡		5.	X				
						6.				
			≡				7.	X	X	
			≡					8.		
									9.	
										10.

The preparation of a matrix will often reveal an unnecessary repetition or redundancy in the list of learning units. In the above example item 10 is tautological, a redefinition or repetition of item 9.

5. The final step is to make a new matrix, reordering the learning units so that the associations and discriminations are brought as close to the diagonal or indirect association chain as is possible. This process often requires considerable heuristic trial or reshuffling in order to get a good fit. The designer should not attempt to get a perfect fit, however, as there may be several close

fits which are equally desirable. An example of a revision of the above learning sequence follows:

Example:

```
1. | X | X | X
   3. | X |   | X
      4.
         5. | X
            8.
               6. | X
                  2. | X | X
                     7. | X
                        9.
```

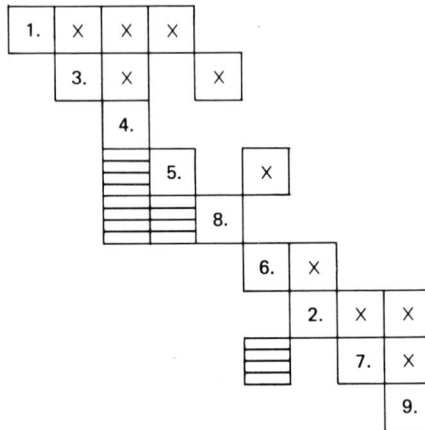

The system concept of equifinality is once again revealed in this matrix process. There are usually several good fits possible once a tentative start and operational limits or goals are determined. The organizer of learning units should not necessarily seek to establish an absolute or perfect linear sequence, but only a close grouping of direct associations and sharp differentiations is productive. However, a little time spent in matrix construction and partial revision often reveals remarkably unsystematic, haphazard, and unproductive learning organization.

The final effect of the matrix technique is to group small learning units into larger units of common or related concepts. In the above example there are obviously three important topic centers—symbols, codes, and maps.

In addition to using matrices for preplanning learning sequences, the instructional programer can validate his program sequences empirically. He can analyze the performance of students in response to each unit or frame in a sequence, employing the 85 percent success criterion as a measure of satisfactory performance. If successful performance response on one item later in the sequence is higher than the response of the previous item the order is questionable. The reverse is also true. However, the method of adjusting these two types of inconsistencies is different. In the first case the prior item may be considered as irrelevant. In the latter case another learning item may be needed to connect or bridge the apparent gap in the learning chain or sequence. If two items in the sequence are responded to unsatisfactorily there is considerable probability that the sequence is incorrect. Poor response to one item only suggests the possibility of sequential error, not the probability.

INSTRUCTIONAL MEDIA SELECTION

Once job/task units have been validated and learning sequence established the next program design problem is the selection of adequate means-methods-media to facilitate the learning transformation. In contrast to the progress made by logical and empirical validation of techniques for determining valid content and optimum sequence, there is little systematic advancement in the knowledge needed for deciding when, where, and how to use different media. Media theory is in a primitive state. There seems to be no feasible way in which media decisions can be empirically validated against performance data. The cost of testing different media under identical conditions tends to restrict development in this area.

If the above assumption is accepted, then the tenets of system analysis suggest that the appropriate medium for a learning task is that medium which can do the job adequately for the least possible cost. Media makes information available; they do not truly select content or organize sequence; they do not determine to any great extent pre- or post-disposition or motivation; they do not teach. The choice of medium seems to be at best a relatively unimportant property of a learning presentation. The medium is but a tool for facilitating the instructional presentation and the learner's response representation.

In view of the present state of knowledge regarding instructional media, it would appear that in an extensive and complex learning program the use of multi-media would tend to enrich the response capabilities of the learner. Certainly one consideration in the selection of media is the determination of the amount of content and pace of instruction that is controlled by the learner. Individual instruction is more expensive than group process. The comparative cost of instructional media for a particular program must consider what materials, equipment, facilities, and tutorial help will be supplied each learner and how these media can be organized to distribute the learning most effectively.

There are some indications that media are generally adaptable for use in the following applicative patterns:

1. Transparencies, training charts, mock-ups, procedure trainers, and simulators facilitate learning identifications, locations.
2. Simulators, part-task trainers, procedures trainers, are useful in learning skilled perceptual-motor acts.

3. Training films, simulators, procedures trainers, are useful in learning procedural sequences or motor chains.
4. Transparencies, simulators, training films assist in learning discriminations.
5. Television, animated panels, training films, operating mock-ups, and simulators are useful in learning concepts, principles, and relationships.
6. Actual equipment, simulators, training films, and teaching machines facilitate problem-solving and trouble-shooting learning.

In spite of these general guidelines, it appears that the best criterion for media selection is to choose the least costly learning medium or media mix which adequately supports the acquisition of the instructional responses.

TESTING AND STANDARDIZING THE PROGRAM DESIGN, COMPLETING PROGRAM DEVELOPMENT

With the selection of appropriate learning media the problem of designing the prototype learning unit is resolved. Before completing the initial design of a prototype learning program, we must account for the following elements:

1. A program or plan of learning tasks, learning responses, cues indicative of learning transmission, and criteria and instruments for measuring response adequacy has been proposed, validated, and standardized via codification.
2. Data supporting learning unit validity is available. This data includes information justifying the learning sequence.
3. A codified, validated, and standardized criterion test is available. Criterion performance or proficiency tests are also built into the program to expedite measurement, motivation, and prediction.
4. A teacher plan or outline of the instructional strategy for the program has been completed which indicates learning objectives, learning content, learning activities, learning media, and measures of adequate performance.
5. An administrative or supervisory plan for putting the prototype program into the laboratory or field for further testing has been completed. This plan should account for a time schedule, the resources assigned to to the job, and the source and number of learners to be involved.
6. All of the above plans or documents have been edited for technical

accuracy, for learning program content or composition, for integration of means-methods-media into the instructional system.

When these tasks have been completed, the production of the prototype learning system is terminated. (Step 9 in the MARS model has been accomplished.)

Following the completion of the instructional program design, it must be given a laboratory tryout. This can be somewhat of an informal tryout at first, often beginning with two or three upper ability learners, persons in the top quartile. Choosing high ability learners first serves several purposes. Through this order of testing, too great a difficulty of content can be immediately determined. (It is more efficient to work down from difficulty parameters than up.) Potential learner difficulties will be readily detected. In addition, the more intelligent learners can point out program weaknesses more readily.

Next the prototype is tested informally with learners selected to represent all abilities—top, medium, and slow students. Finally, the laboratory test involves a large or normative group tryout, evaluation, and revision. MARS Step 10 is then completed.

At this time field testing begins. The program is placed in operation in facilities and conditions representative of the environment it will experience in normal production. It is tested and evaluated in the field; any revisions are incorporated into the prototype and given a full field test also. Logs are kept by instructors and supervisors in the field. Critical incidents related to program operations and learner responses are systematically recorded. Suggestions for incorporating materials, media aids, etc. and improving administrative methods are recorded and reported. MARS Step 11 is now terminated.

When a full field test of the model program has proved to be satisfactory, preparations are made for full-scale development and dissemination of the program. Steps 12–14 of the MARS model are put into operation. Conservative and realistic (normative) standards for learners and instructors are established. Revisions of criterion tests are made to insure their compatibility with any revisions in the program. Instructional manuals are updated.

Administrative specifications for facilities, equipment, instructional organization, general support communications and logistics are established. Materials, machines, media tools and aids are purchased or produced.

A critical step in development is the thorough training of personnel needed to operate the program. Another important task in development involves the dissemination of news of the program to appropriate target audiences.

Before the instructional materials are distributed, a careful evaluation and editing of the instructor's manuals is completed. The revised manuals should include improved course descriptions, target learner population descriptions, behavioral response learning objectives, criterion tests of the desired types, norm-referenced performance tests if necessary, system performance data, information on administration, suggestions for motivating and counseling, suggestions for enrichment and for adjusting the program to fast learners or to those who enter late. A sample course of study or learning unit is included in the instructional manual.

All program materials and resources are then allocated and inventoried. The machines and technology are set up and tested in trial runs. Personnel are field tested and evaluated. The program is now ready for full-scale operation. MARS Step 14 is then complete.

Finally the program is placed in full-scale operation. If thoroughly prepared, it is carefully and systematically designed to facilitate the specific learning reflected in its objectives. The instructor now becomes an administrator or manager of learning resources, not a dominant resource. He can now play a larger role in counseling, motivating, tutoring, or diagnosing.

In administering a carefully designed learning program, special provisions need to be made to accommodate those learners who finish early and those who enter late. Ideally the larger environmental system into which the program is introduced encourages flexibility, continuous progress, and individualization. In a well-designed programed learning unit flexible self-pacing is highly desirable.

The final step and subprogram in the overall cycle of instructional system analysis and program design is now operative. The management control subsystems are revised to coordinate and articulate the new program with the educational system as a whole. Administrative attention and related decisions now are directed toward maintenance and support and toward continuous operational evaluation and revision of subelements and of the system as a whole. If carefully executed, the model planning procedure assures a high degree of probable success, for it has systematically and sequentially resolved uncertainties regarding functional possibility, technological adequacy, economic feasibility, and social-political desirability.

22

A Systematic Philosophy of Learning and Living

It is almost inevitable that persistent and accumulating human inquiry and knowledge have constantly extended human learning and living in the direction of higher levels of abstraction and into the domain of philosophy. The true scientist or theorist soon discovers that he must articulate and coordinate his disciplinary knowledge into an interdisciplinary or metatheoretical perspective and style of life. He must resolve all of the inconsistencies brought about by an overspecialized, goal-displaced education and mode of living which, for practical reasons, separates man and nature, human judgment and human technology, the inquiry modes of history and science, learning and living, human esthetic-ethical values and human practical-operational values.

Thus the method of inductive-deductive thought has repeatedly generated in the minds of the great systematic thinkers of the past two hundred years an empirical human esthetic-ethical theory for living as a necessary enlargement of the rules and methods of science and systems, one which accommodates human imagination. The system perspective must generalize such a system of human values. It cannot logically permit the subtraction of this human component from the model of the universe and expect the model of the part to satisfactorily represent the whole itself.

So systematic philosophy has generated, like dogmatic and doctrinaire philosophies, attempts to account for universals, but its universals are dynamic as are nature and human behavior. Yet systematic philosophy, like the purely relativistic philosophies of Sartrean and McLuhan-type intuitivism and immediacy, accounts for particulars and events, but its events are conditionally relative and not totally anarchistic and anomic.

There is a very definitive structure of order and hierarchy in the systematic or scientific perspective. And the eclectic systematic philosopher integrates historical esthetics and ethics with scientific and logical procedures, producing an internally consistent whole.

Founded in inductive-deductive inquiry, the systematic perspective requires that an adequate human philosophy must be both created and discovered by man. A part of its structure must be initiated inspirationally or imaginatively by individuals, socially communicated, tested in human experience, recorded in art, history, and scientific inquiry, and finally subjected to persisting and organismic reflective thought. Thus it cannot possibly conflict for long with any substantive experiences of reality or of inquiry. And it is not entirely dependent upon, reactive to, or justified by nature alone (the mere probability of that which is), for human imagination has been always quite independent of past and present environmental restrictions.

Alexander Sesonske's book *Value and Obligation* is an excellent beginning reference for the educator who wishes to contemplate the basic elements and procedures for understanding the idea of an empirical social ethic. He explains so well the ethical necessity of human interaction, mutuality and freedom of choice in any qualitative social commitment.

A systematic philosophy has a degree of openness or tentativeness in its structure. It provides for human imagination and belief, for individual creation and discovery that can be reconstructed into social process and utopian vision. It accounts for the possibility of faith, hope, and freedom of human will—thus balancing human imagination and invention with discovery and prediction. Within the system frame of reference, human control of systems may transcend events in successive ways, demonstrating the possibility of (but not the necessary and sufficient conditions for) being positive, integrating, and organismic. The succession of human and natural events within a totally absolutist, closed, naturally scientific, or reconstructive perspective is merely probable and amoral, without any choice or responsibility for human long-range direction and integration.

Traditional naive philosophies and doctrines which have sustained western culture for many generations were weakened considerably by the rise of science and technology. The events of automation and the generation of related disciplinary and interdisciplinary theoretical systems have further accelerated the scholar's awareness of changing human values and systems of values.

The secularization and professionalization of human inquiry have irreversibly spread until they permeate all facets of formal education, facilitating the destruction of dogma and doctrine as the base for educational purposes and processes. Even Communism has undergone systematic transformations during its relatively short life, for it must constantly and inevitably subject itself to historical and scientific scrutiny and judgment.

Education in the United States, representing both traditional and emerging national normative characteristics and processes, has retained its content of naive idealism and optimistic utopian vision to some degree. True democracy, freedom of inquiry, expression, and choice, and the extended sharing of human rights among all peoples are replacing a declining authoritarianism and a professed ideology of absolutist free enterprise (the rights of the possessor and survival of the fittest dogma). The democratization and education of the masses in America are quite responsible for both the amazing success and obvious limitations of our social-political and our scientific-technological ways of life.

In recent years our democracy, in conjunction with and in philosophical and operational conflict with our technology of machines and bureaucratic organizations, has experienced the complications of persisting success in both areas. Inquiry and technology have reduced social and environmental coercion and threat simultaneously, although our pragmatism has directed our major social efforts primarily in the direction of a narrow utilitarian economics and material production. The lay political control of schools and the traditionally specialized rather than liberal training of teachers has delayed the dissemination of emerging dynamic and systematic philosophies. Where political dogma and authoritarianism have been effectively eroded, religious dogma has often slowed the progress of a more integrated cultural change.

National political and educational efforts in interdisciplinary curriculum development have been necessary to bring about the widespread and significant changes in mathematics, the natural sciences, linguistics and semantics, and in the social disciplines. Their combined import and significance is only understood by the intellectual elite at present, but the inevitable and irreversible pattern of human inquiry and information flow will lead to greater dissemination.

We are now beginning to teach comparative political and religious doctrines in our more progressive high schools as well as in our universities and colleges. There is no surer way of destroying or modifying doctrine and dogma than this. For, exposed to free inquiry, any dogma or doctrine will be discovered to be perhaps a historically useful and even remarkably beautiful metaphor or model for living, but, as a model, it will eventually be understood to be incomplete—if not inconsistent. Our emerging models of science and systems integrated into an empirical social ethic will, by comparison, display a more dynamic adaptability, more internal and external consistency, more utility, and a more organismic nature.

As American educational administrators and teachers become more thoroughly and liberally educated, they will begin to appreciate and reflect the value of a truly integrated and liberal education as never before—present trends toward academic specialization notwithstanding. They will realize the impossibility and impracticality of facilitating basic economic and welfare values through technology and denying the advancement of political and esthetic values at the same time. They will recognize also that it is not only fear and ignorance which has delayed the emergence of social justice and esthetics, but strategic and intelligent efforts to maintain the deferent status and power of certain selfish factions in our nation (and in all other nations, of course).

Our present liberally educated students are learning through the exercise of inquiry that the truly great men of history and of science have been systematically and progressively freed from the restrictions of particularistic doctrines and activities of expediency, however. Immanual Kant and Ashley Montagu have previously been quoted, and their personally and reflectively generated versions of the Golden Rule have been presented. These metaphors exemplify an operational rule of living that is both necessary and desirable for the future happiness and welfare of mankind. Applied to organizations and societies these value statements can be interpreted as an inferential base for a belief in social democracy, equality, justice, freedom, human cooperation, and concern.

John Dewey described the educational and personal philosophy which lacks either esthetic and ethical idealism or natural and practical realism as follows:

> Idealism easily becomes a sanction of waste and carelessness; and realism a sanction of legal formalism in behalf of things as they are— the rights of the possessor.
> (Dewey, *On Experience, Nature, and Freedom*, p. 69)

He wrote that any theory of activity in social and moral matters which is not grounded in a comprehensive and integrated philosophy is but a projection of arbitrary personal preference. Dewey further commented:

> Success, power, freedom in special fields is in a maximum degree relatively at the mercy of external conditions . . . But against kindness and justice there is no law.
>
> (ibid., p. 278)

Albert Einstein expressed his philosophy for living in the words, "Only a life lived for others is a life worthwhile." He also expressed a belief that man cannot experience freedom unless he can experience the dual freedoms of arbitrary decisions and moral responsibility. Einstein believed that there is no morality or ethical responsibility for men without free will, that there is no personal or social ethic without human purpose and human control.

The writings of other great men of science and philosophy of the last two centuries and particularly of the last two decades reflect this same theme. They profess a common social ethic, optimistic and esthetic in character, and rather simply expressed.

Educators who profess to educate or administer without an integrating and self-motivating cognized, systematized, and exercised philosophy cannot prevent being mere reactors to their past experiences, to social tradition, or to their natural environment. They are in no sense social leaders or statesmen. Utilitarianism is not characteristic of great educators any more than of great scientists and philosophers. The world of education as well as the world of science and philosophy has attained a new dimension in the light of post-Einstein thought.

For the educational administrator who wishes to assume an ethically-based professional leadership responsibility, further reading in all modern systematic philosophies and in all modern theory-based disciplines is recommended. The system thread permeates them all.

Educators who believe that former social ideologies based on authoritarian dogma and doctrine will return to dominate education in the future probably are deluding few but themselves. This cannot occur in the ultimate without the complete control of human inquiry and information, and the control of either on a world scale is not only difficult but probably impossible. History tells us that inquiry and education fundamentally are irreversible processes. The power of our modern speculative philosophies, systematized theories, and socialized technologies is widely distributed and is becoming more widely disseminated with each passing day. Any modern nation that restricts inquiry and technology in some arbitrary and unbalancing manner will fall from within or without.

This does not mean that American educators do not need to work to preserve and extend their democracy, freedom, and the social ethic, however. The very opposite is true. Democracy and justice, to exist, must be practiced. Voluntary mutual cooperation cannot be developed by nor is it dependent upon coercive and competitive leadership.

As for the educator-technician, he must either formulate some sort of integrated ethical philosophy and style of life or become the pawn of those that do. There is little doubt that the lack of an intellectually acceptable general philosophy of education partially accounts for the minor cultural influence traditional schools have had and are having on American social and economic values.

It is intended and hoped that professional educators will not interpret these professorial comments to be critical or pessimistic in vein. The fact is, that among all of America's institutions, the schools singly exemplify and demonstrate the persistence and current existence of an active and at least partially formulated and practiced dynamic social ethic. Educators and educational institutions, often regarded as simplistic and naive, may actually be closer to truth (necessary and sufficient significance and consistency) in their words and actions than most individuals and organizations in American society. Their naive utopian vision combined with an emerging stress on free inquiry may eventually integrate into a synergistic whole the myriad of individual successes and events our suboptimized technology, science, and systems have made possible. For, as man has the power to imagine, he has the power to believe, to inquire, to know, to interpret, and to value. As he has the power to value he has the power to transform, to improve upon present conditions. To be maximally productive, personally and socially, and to live a satisfying life, men must educate themselves and others to develop attitudes and behaviors which integrate ideals and efforts. The philosophy of system accommodates and facilitates such ends-means relationships.

Yet men will never reach a stage where living is a science rather than an art. The great natural scientists and theorists are the first to acknowledge the complexity in nature as revealed in our great scientific laws and theories. Einstein has said that the comprehensibility of the world is eternally incomprehensible. This is as it should be for man and as it will be in nature. Only an empirical, esthetic, and ethical system of social and environmental relationships will prevail if man is to survive in the process of the evolution of nature and human imagination.

BIBLIOGRAPHY

Bibliographical References and Primary Sources

AASA (American Association of School Administrators). *Administrative Technology and the School Executive.* Washington, D.C.: AASA, 1969.

Abt, Clark. *Games for Learning.* Cambridge, Mass.: Occasional Paper Number 7, Educational Services, Inc., 1966.

Ashby, W. R. *Introduction to Cybernetics.* New York: Wiley, 1956.

Ayer, Alfred. *Language, Truth, and Logic.* New York: Dover, 1946.

Beer, Stafford. *Cybernetics and Management.* New York: Wiley, 1959.

Berelson, Bernard, and Gary Steiner. *Human Behavior: An Inventory of Scientific Findings.* New York: Harcourt, 1964.

Bertalanffy, Ludwig von. "General Systems Theory." *Yearbook for the Society for the Advancement of General Systems Theory,* Ann Arbor, 1955.

Bloom, Benjamin (editor), et al. *Taxonomy of Educational Objectives. Handbook I: Cognitive Domain.* New York: McKay, 1956.

Borko, Harold (editor), et al. *Computer Applications in the Behavioral Sciences.* Englewood Cliffs: Prentice-Hall, 1962.

Buckley, Walter (editor). *Modern Systems Research for the Behavioral Scientist.* Chicago: Aldine, 1968.

Burck, Gilbert. *The Computer Age and Its Potential for Management.* New York: Harper, 1965.

Burke, John G. (editor). *The New Technology and Human Values.* Belmont, Calif.: Wadsworth, 1968.

Bursk, E. C., and J. F. Chapman (editors). *New Decision-Making Tools for Managers.* New York: New American Library, 1963.

Campbell, J. H., and H. W. Hepler (editors). *Dimensions in Communications.* New York: Appleton-Century-Crofts, 1960.

Carpenter, F., and E. Haddan. *Systematic Applications of Psychology to Education.* New York: Macmillan, 1964.

Carver, F. D., and T. J. Sergiovanni (editors). *Organization and Human Behavior.* New York: McGraw-Hill, 1969.

Cohen, R. R., and E. Nagel. *An Introduction to Logic.* New York: Harcourt, 1962.

Condon, John C., Jr. *Semantics and Communication.* New York: Macmillan, 1966.

Cook, Desmond L. *Program Evaluation and Review Technique: Applications in Education.* Monograph No. 17, Washington, D.C.: U.S. Government Printing Office, 1966.

Cooper, W. W., H. J. Leavitt, and M. W. Shelley II (editors). *New Perspectives in Organizational Research.* New York: Wiley, 1962.

Daedalus. Toward the Year 2000. Cambridge, Mass.: American Academy of Arts and Sciences, 1967 (Summer).

Dale, Ernest (editor). *Readings in Management.* New York: McGraw-Hill, 1965.

De Cecco, John (editor). *The Psychology of Language, Thought, and Instruction.* New York: Holt, 1967.

Dechert, Charles R. (editor). *The Social Impact of Cybernetics.* New York: Simon and Schuster, 1966.

Dewey, John. *Essays in Experimental Logic.* New York: Dover, 1966.

Dewey, John. *Experience and Education.* New York: Macmillan, 1938.

Dewey, John. *How We Think.* Boston: Heath, 1910.

Dewey, John. *On Experience, Nature, and Freedom.* New York: Bobbs-Merrill, 1960.

Dodd, Stuart C. "In Classification of Human Values: A Step in Predicting Human Valuing," *American Sociological Review,* vol. 16, 1951.

Duncan, Guy. "What Do Americans Value?" *Educational Leadership,* Vol. 20, May, 1963.

Elam, Stanley. *Education and the Structure of Knowledge.* Chicago: Rand McNally, 1964.

Etzioni, Amitai. *Modern Organizations.* Englewood Cliffs: Prentice-Hall, 1964.

Evans, L. H., and G. E. Arnstein. *Automation and the Challenge to Education.* Washington, D.C.: National Education Association, 1962.

Fisk, Milton. *A Modern Formal Logic.* Englewood Cliffs: Prentice-Hall, 1964.

Fuller, Buckminster. "Report on the Geosocial Revolution," *Saturday Review.* vol. 150, no. 36, September 16, 1967.

Gagné, Robert M. "The Learning Requirements for Inquiry." *Journal of Research in Science Teaching,* vol. 1, no. 2, 1963.

Gallagher, J. D. *Management Information Systems and the Computer.* New York: American Management Association, 1961.

Getzels, Jacob W. "The Acquisition of Values in School and Society," *The High School in a New Era.* Chicago: University of Chicago Press, 1958.

Good, Carter, and D. E. Scates. *Methods of Research.* New York: Appleton-Century-Crofts, 1954.

Graff, O. B., C. M. Street, and A. R. Dykes. *Philosophic Theory and Practice in Educational Administration.* Belmont, Calif.: Wadsworth, 1966.

Gue, R. L., and M. E. Thomas. *Mathematical Methods in Operations Research.* New York: Macmillan, 1968.

Guilbaud, G. T. *What is Cybernetics?* New York: Criterion, 1959.

Guilford, J. P. *Personality.* New York: McGraw-Hill, 1959.

Haga, Enoch. *Understanding Automation.* Elmhurst, Ill.: The Business Press, 1965.

Hall, Robert A. Jr. *Linguistics and Your Language.* Garden City: Doubleday, 1960.

Halpin, Andrew W. *Theory and Research in Administration.* New York: Macmillan, 1966.

Hempel, Carl, et al. *Frontiers of Science and Philosophy.* Pittsburgh: University of Pittsburgh Press, 1962.

Herberg, T., and J. Bristol. *Elementary Mathematical Analysis.* Boston: Heath, 1962.

Hilton, Alice Mary. *Logic, Computing Machines, and Automation.* Cleveland: World, 1964.

Hoffer, Eric. *The Ordeal of Change.* New York: Harper, 1952.

International Business Machines. *Data Processing Techniques Form C 20–8152.* White Plains, New York: IBM Technical Publications Department.

Kahn, H. and A. Wiener. *The Year 2000.* New York: Macmillan, 1967.

Kaimann, R. A. "Educators and PERT." *Journal of Educational Data Processing,* vol 3, no. 2, Spring, 1966. Malibu, Calif.: Educational Systems Corp.

Kaplan, Abraham. *The Conduct of Inquiry: Methodology for Behavioral Science.* Scranton: Chandler, 1964.

Kardiner, A., et al. *The Psychological Frontiers of Society.* New York: Columbia University Press, 1945.

Kerlinger, Fred N. *Foundations of Behavioral Research.* New York: Holt, 1966.

Korzybski, Alfred. *Science and Sanity,* fourth edition. Lakeville, Conn.: The International Non-Aristotelian Library, 1958.

Krathwohl, D. R. et al. *Taxonomy of Educational Objectives. Handbook II: Affective Domain.* New York: McKay, 1964.

Lasswell, H. D., and A. Kaplan. *Power and Society.* New Haven: Yale University Press, 1950.

Lewis, C. I., and C. H. Langford. *Symbolic Logic,* second edition. New York: Dover, 1959.

Lippitt, R., J. Weston and B. Westley. *The Dynamics of Planned Change.* New York: Harcourt, 1958.

Lipschutz, Seymour. *Outline of Theory and Problems of Set Theory and Related Topics.* New York: Schaum Publishing Co., 1964.

Loughary, John W. et al. *Man-Machine Systems in Education.* New York: Harper, 1966.

Lyden, F. J., and E. G. Miller (editors). *Planning Programing Budgeting.* Chicago: Markham Publishing Co., 1968.

McCarthy, John, et al. "Information." *Scientific American,* vol. 215, No. 3, September, 1966.

McLuhan, Marshall. *Understanding Media: The Extensions of Man.* New York: McGraw-Hill, 1965.

Mansfield, Edwin (editor). *Managerial Economics and Operations Research.* New York: Norton, 1966.

March, James G., and Herbert A. Simon. *Organizations.* New York: Wiley, 1958.

Marks, Robert W. *Simplifying Set Theory*. New York: Bantam, 1966.

Maslow, A. H. *Eusychian Management*. Homewood, Ill. Irwin, 1965.

Maslow, A. H. *Motivation and Personality*. New York: Harper, 1954.

Merleau-Ponty. *Phenomenology of Perception*. New York: Humanities Press, 1962.

Miller, D. W., and M. K. Starr. *The Structure of Human Decisions*. Englewood Cliffs: Prentice-Hall, 1967.

Mills, C. Wright. *The Sociological Imagination*. London: Oxford University Press, 1959.

Montagu, Ashley. *The Humanization of Man*. New York: Grove, 1962.

Morphet, E. L., and D. L. Jesser (editors). *Planning for Effective Utilization of Technology in Education*. New York: Citation Press, 1969.

Morris, Charles. *Signs, Language, and Behavior*. Englewood Cliffs: Prentice-Hall, 1946.

Muessig, Raymond H. (editor). *Youth Education*. Washington, D.C.: ASCD 1968 Yearbook, National Education Association, 1968.

Nadler, Gerald. *Work Systems Design: The Ideals Concept*. Homewood, Ill.: Irwin, 1967.

Nagel, Ernest. *The Structure of Science: Problems in the Logic of Scientific Explanation*. New York: Harcourt, 1961.

NSSE Yearbook. *Behavioral Science and Educational Administration, Part II*. Chicago: University of Chicago Press, 1964.

Oettinger, G., and S. Marks. *Run Computer, Run*. Cambridge, Mass.: Harvard University Press, 1969.

Office of Economic Opportunity. *Instructional Systems Development Manual*. Washington, D.C.: JCH 440.4., March, 1968.

Ortega y Gasset, Jose. *History as a System*. New York: Norton, 1961.

Pask, Gordon. *An Approach to Cybernetics*. New York: Harper, 1961.

Peirce, Charles S. *Selected Writings: Values in a Universe of Chance*. New York: Dover, 1958.

Perloff, Harvey S. (editor). *Planning and the Urban Community*. Pittsburgh, Pa.: University of Pittsburgh Press, 1967.

Pfiffner, J. M., and F. P. Sherwood. *Administrative Organization*. Englewood Cliffs: Prentice-Hall, 1960.

Phillips, Bernard S. *Social Research: Strategy and Tactics*. New York: Macmillan, 1966.

Pigors, P.; C. Myers, and F. Malm (editors). *Management of Human Resources.* New York: McGraw-Hill, 1964.

Polanyi, E., *Personal Knowledge.* Chicago: University of Chicago Press, 1958.

Porter, Elias H. "The Parable of the Spindle." *Harvard Business Review,* May–June, 1962.

Presthus, Robert. *The Organizational Society.* New York: Knopf, 1962.

Raiffa, Howard. *Decision Analysis.* Reading, Mass.: Addison-Wesley, 1968.

Rapoport, Anatol. *Two-Person Game Theory: The Essential Ideas.* Ann Arbor: University of Michigan Press, 1966.

Read, Herbert. *The Forms of Things Unknown.* Cleveland: World, 1963.

Rudner, Richard S. *Philosophy of Social Science.* Englewood Cliffs: Prentice-Hall, 1966.

Russell, Bertrand. *A History of Western Philosophy.* New York: Simon and Schuster, 1945.

Russell, Bertrand. *Human Knowledge, Its Scope and Limits.* New York: Simon and Schuster, 1948.

Russell, Bertrand. *The Problems of Philosophy.* New York: Oxford University Press, 1959.

Ryans, David G. *An Information Systems Approach to Education.* Santa Monica, Calif.: TM-1945, Systems Development Corporation, 1963.

Scheffler, Israel. *Conditions of Knowledge.* Chicago: Scott, Foresman, 1965.

Schilpp, Paul A. (editor). *Albert Einstein: Philosopher-Scientist, volume II.* New York: Harper, 1959.

Sesonske, Alexander. *Value and Obligation.* New York: Oxford University Press, 1964.

Simon, Herbert A. *The New Science of Management.* New York: Harper, 1960.

Smith, K. U., and M. F. Smith. *Cybernetic Principles of Learning and Educational Design.* New York: Holt, 1966.

Sondel, Bess. *The Humanity of Words.* Cleveland: World, 1958.

Strauss, Erwin. *Phenomenological Psychology.* New York: Basic Books, 1967.

Taba, Hilda. *Curriculum Development: Theory and Practice.* Englewood Cliffs: Prentice-Hall, 1960.

Taube, Mortimer. *Computers and Common Sense.* New York: Columbia University Press, 1961.

Thelen, Herbert A. *Dynamics of Groups at Work.* Chicago: University of Chicago Press, 1954.

Thurman, Kelley (editor). *Semantics.* Boston: Houghton Mifflin, 1960.

Tillitt, H. E. (writing chairman). *Computer Oriented Mathematics.* Washington, D.C.: National Council of Teachers of Mathematics, 1963.

Toulmin, Stephen. *Foresight and Understanding.* Bloomington: Indiana University Press, 1960.

Van Dalen, D. B., and W. J. Meyer. *Understanding Educational Research.* New York: McGraw-Hill, 1962.

Whitehead, A. N., and Bertrand Russell. *Principia Mathematica.* Cambridge: The University Press, 1964.

Wiener, Norbert. *Cybernetics.* New York: Wiley, 1961.

Wiener, Norbert. *The Human Use of Human Beings.* New York: Avon, 1964.

Index

371